Dialectical Passions

COLUMBIA THEMES IN PHILOSOPHY,
SOCIAL CRITICISM, AND THE ARTS

COLUMBIA THEMES IN PHILOSOPHY, SOCIAL CRITICISM, AND THE ARTS
Lydia Goehr and Gregg M. Horowitz, Editors

Columbia Themes in Philosophy, Social Criticism, and the Arts presents monographs, essay collections, and short books on philosophy and aesthetic theory. It aims to publish books that show the ability of the arts to stimulate critical reflection on modern and contemporary social, political, and cultural life. Art is not now, if it ever was, a realm of human activity independent of the complex realities of social organization and change, political authority and antagonism, cultural domination and resistance. The possibilities of critical thought embedded in the arts are most fruitfully expressed when addressed to readers across the various fields of social and humanistic inquiry. The idea of philosophy in the series title ought to be understood, therefore, to embrace forms of discussion that begin where mere academic expertise exhausts itself, where the rules of social, political, and cultural practice are both affirmed and challenged, and where new thinking takes place. The series does not privilege any particular art, nor does it ask for the arts to be mutually isolated. The series encourages writing from the many fields of thoughtful and critical inquiry.

Lydia Goehr and Daniel Herwitz, eds., *The Don Giovanni Moment: Essays on the Legacy of an Opera*
Robert Hullot-Kentor, *Things Beyond Resemblance: Collected Essays on Theodor W. Adorno*
Gianni Vattimo, *Art's Claim to Truth*, edited by Santiago Zabala, translated by Luca D'Isanto
John T. Hamilton, *Music, Madness, and the Unworking of Language*
Stefan Jonsson, *A Brief History of the Masses: Three Revolutions*
Richard Eldridge, *Life, Literature, and Modernity*
Janet Wolff, *The Aesthetics of Uncertainty*
Lydia Goehr, *Elective Affinities: Musical Essays on the History of Aesthetic Theory*
Christoph Menke, *Tragic Play: Irony and Theater from Sophocles to Beckett*, translated by James Phillips
György Lukács, *Soul and Form*, translated by Anna Bostock and edited by John T. Sanders and Katie Terezakis with an introduction by Judith Butler
Joseph Margolis, *The Cultural Space of the Arts and the Infelicities of Reductionism*
Herbert Molderings, *Art as Experiment: Duchamp and the Aesthetics of Chance, Creativity, and Convention*
Whitney Davis, *Queer Beauty: Sexuality and Aesthetics from Winckelmann to Freud and Beyond*

Dialectical Passions

Negation in Postwar Art Theory

Gail Day

Columbia University Press
New York

Columbia University Press
Publishers Since 1893
New York. Chichester, West Sussex

Library of Congress Cataloging-in-Publication Data
Day, Gail.
Dialectical passions : negation in postwar art theory / Gail Day.
p. cm. — (Columbia themes in philosophy, social criticism, and the arts)
Includes bibliographical references and index.
ISBN 978-0-231-14938-9 (cloth : alk. paper) — ISBN 978-0-231-52062-1 (electronic)
1. Art, Modern—20th century—Philosophy. 2. Art, Modern—21st century—Philosophy. 3. Negation (Logic) I. Title. II. Series.
N6490.D34 2011
701'.18—dc22
2010004988

Contents

Illustrations

Acknowledgments

My most important debt is to Steve Edwards for his unflagging patience, commitment, and engagement. I doubt *Dialectical Passions* would have been completed without his personal support and his astute readings of its countless drafts. I have also been especially fortunate to have received the inspiration, guidance, and friendship of Fred Orton and wish to take this opportunity to say how thankful I am. Caroline Arscott, John X. Berger, Andrew Hemingway, Alex Potts, Susan Siegfried, and Julia Welbourne have been sustaining presences through the gestation of this book, providing critical insights into my ideas and offering crucial encouragement to persist when stasis prevailed. To the readers of the manuscript, in part or in whole, and at various stages of its development, I owe particular thanks: Caroline (again), Martin Gaughan, Ken Hay, Stewart Martin, Adrian Rifkin, John Roberts, and Fred Schwartz. Their advice and incisive comments have helped clarify and develop my arguments. Along the way I have also benefited from discussion and debate with Joanne Crawford, Tim Hall, Catherine Lupton, David Mabb, April Masten, Stanley Mitchell, Jo Morra, Ben Noys, Peter Osborne, Giles Peaker, Chris Riding, Nick Ridout, Frances Stracey, Nick Till, and Ben Watson. Invitations to present work-in-progress from Tim Clark, David Cunningham, John Goodbun, Tom Hickey, Neil Leech, Colin Mooers, and Julian Stallabrass, and also from Alex and Andrew, have enabled earlier versions of my research to be tested in sympathetic, yet challenging environments.

For their belief in the project, I am grateful to my series editors Lydia Goehr and Gregg Horowitz and to Wendy Lochner at Columbia University Press. My appreciation also goes to Christine Mortlock and Leslie Kriesel for ensuring its realization and particularly to Robert Demke for his essential guidance and his help in improving my manuscript.

Immense gratitude goes to my family—especially to Jean Day, Brian Day, and Edna Colledge—for enduring all the uncertainties and long absences while I worked on this book.

The Arts and Humanities Research Board, Wimbledon School of Art, and the University of Leeds provided generous support to enable me to focus on writing, and, in addition, the University of Leeds has helped toward the costs of images and reproduction rights. Less formally, but no less importantly, the unique combination of peace, sanity, and hospitality at La Touche has been invaluable for securing periods of concentration necessary for writing and thinking. Earlier and shorter versions of chapters have appeared in *Oxford Art Journal* 22, no.1 (1999): 103–118; *Art History* 23, no. 1 (2000): 1–18; and *Radical Philosophy* 133 (2005): 26–38.

Dialectical Passions

Introduction

Amid reflections on the rituals, commodity kitsch, and everyday banalities of modern Japanese life, Chris Marker, in his film *Sans Soleil*, turns his attention to a Left-wing demonstration. The camera alights briefly on a middle-aged man (see figure 0.1). The scene is shot at a rally to commemorate the birthday of a victim of a protest at the same site in the nineteen-sixties, when peasants fought to prevent an airport being built on their land. The repetitions and echoes between "then" and "now" make the occasion "unreal," part of the "world of appearances." The man—who we take to be one of the displaced farmers—cuts a solitary figure among his youthful comrades. The airport, of course, was built. With his withering comments on the utopias of the Left, and the descent of so many of its student participants into "postures and careerists," Marker certainly prods his viewers toward the melancholic conclusion that the future is already written in this experience. But the lure—one of many that shape the film—serves to underscore Marker's recalcitrance: he has in mind neither "resignation" nor "defeat," but rather a reminder of how human subjects are fundamentally transformed in and through the processes of resistance. The airport is "more besieged than victorious." He may be on the losing side, but once touched by the awareness and experience that comes through political protest, this man's outlook could never again be what it was.

> I met peasants there who had come to know themselves through the struggle. Concretely it had failed. At the same time, all they

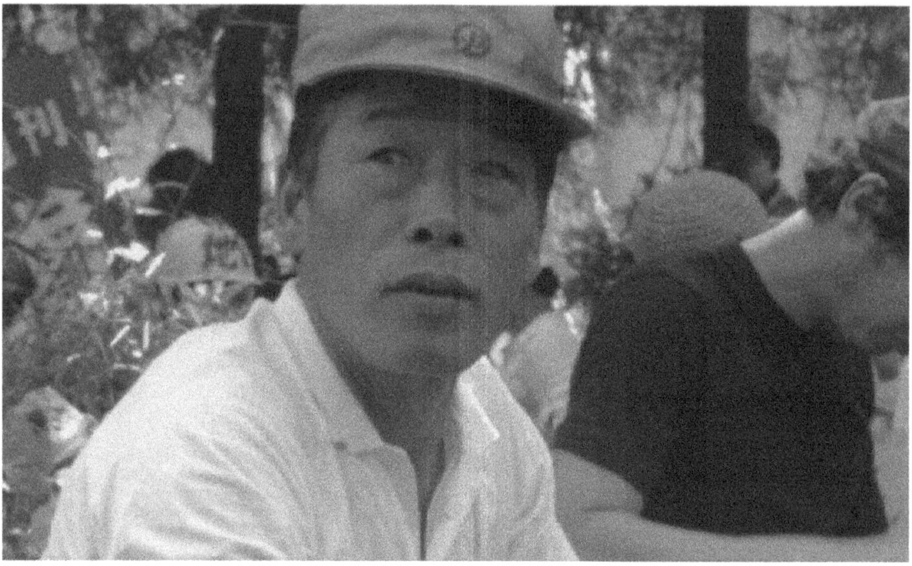

0.1 Chris Marker, film still from *Sans soleil* [*Sunless*], Argos Films, 1982. Courtesy of the artist.

> had won in their understanding of the world could have been won only through the struggle.[1]

A century or so earlier, another radical cultural worker reflected on the problem of defeat and agency:

> men and women fight and lose the battle, and the thing that they fought for comes about in spite of their defeat, and when it comes turns out not to be what they meant, and other men and women have to fight for what they meant under another name.[2]

I have taken the liberty of extending the gender reference of William Morris's echo of Hegel's ruse of reason. Cited by E. P. Thompson and Perry Anderson, among others, this passage also serves as epigraph to Michael Hardt and Antonio Negri's *Empire*.[3] The lines are framed by the narrator's distracted ruminations. Listening to John Ball's speech, his mind wanders (although not entirely) and—we are twice reminded—he ponders how struggles are recuperated and have to be renewed. Marker too makes an art of this radical rumination, reflective and critical by turns, that the term "essayistic"—often used to describe his filmmaking—fails adequately to capture.

My study considers certain radical trajectories from discussions on art and architecture. Taking as its object the field of the history, theory, and criticism of art in the latter half of the twentieth century, *Dialectical Passions* pursues the critical reflections on the persistence and renewal of resistance. The title calls attention to the heated debates, fractures, and critical legacies emerging from the New Left. More specifically, the chapters that follow attend to arguments staked in and around dialectical thought and, most especially, in and around the dialectics of social process.[4] This perspective is unusual, not just for its primary focus on art and architecture (literature is more usually treated as the privileged site of cultural theory) but also for its critical perspective on some of the dominant modes of cultural theory that came to prominence over the last forty years or so. Addressing themes such as critical distance, mediation, and totality, and the work of thinkers who engaged substantially with these problems, this book tracks lineages that—while by no means marginal—stand somewhat apart from the main directions taken by most overviews of art theory. An overarching theme here is the varying attitude toward negation, and it is an attention to some of the valences of negative thought in contemporary art theory that gives this book its particular focus. I attempt neither to map the politics of the discipline of art and architectural history, nor specifically to trace politics in it. Addressing debates on symbol and allegory, or those articulated around the concepts of spectacle, metropolis, or avant-garde, this book explores the space between a radicalized approach to dialectics and Left-orientated nihilism. This focus delimits the field primarily to those writers whose work stakes political claims, whose writing engages with varying conceptions of emancipatory thought, or opens historically and methodologically reflexive approaches pertinent to (if not always consonant with) dialectics. I consider the contributions of some leading figures in late-twentieth-century art theory, such as Benjamin Buchloh, Hal Foster, and Fredric Jameson—the latter especially with regards to his writing on architecture. Accounts presented by significant figures in art history, such as Rosalind Krauss, Craig Owens, Douglas Crimp, and Michael Fried, also play a role. The writings of two thinkers—T. J. Clark and Manfredo Tafuri—provide the basis for extended meditations. Their sustained engagements with the politics of dialectics—rooted, in the one instance, in the Left-Hegelian militancy of the Situationist International, and, in the other, the anti-Hegelian vision developed within Italian workerism—represent approaches that have not been readily assimilated into the wider intellectual consensus. Despite occupying a canonical position in the histories of art and architecture respectively, their disciplines have yet to come to terms with this aspect of their interventions. Both Clark and Tafuri pursue issues arising from negative thought, and a process of dialectical unfolding is worked out in their accounts of art—not in the systematic manner of professional philosophers, but in ways that connect (in my view, far more interestingly) with its "pulse of freedom."[5] In the light of the recent resurgence of radical discourse, and the efforts to reconsider and historicize the projects of the New Left, these projects assume fresh relevance. Their work provides the focus of the first two chapters. In chapter 3, I consider the debate on allegory, starting out from its revival in the nineteen-eighties as a framework for understanding a critical form of postmodernism (advanced, above all, by Craig Owens) and addressing its implications for dialectical thinking. The final chapter is organized around arguments concerning "social abstraction," a theme that underlies the preceding sections. I trace how this idea, developing out of responses to Georg Lukács's notion of reification (from Theodor Adorno and Max Horkheimer to Guy Debord), becomes the backbone for socially inclined and critical art theory in the work of Buchloh, Foster, and Jameson.

Proceeding through detailed readings, and attending to the specificity of historical and critical texts, arguments are followed in which the object of study provides the dynamic of critical reflection—an approach that provides a means to negotiate some of the complex terrain of negation in the art theory, but which is also true to its object. This commitment has come at a cost, limiting the number of studies that could be included (I am all too aware, for instance, of the absence of an analysis of Peter Bürger or the German New Left); it also means surrendering some of the sureties of external critique. While I would not go as far as Adorno when he claims that "Transcendent criticism sympathizes with authority in its very form, even before expressing any content," I concur with his riposte to those who would dismiss an alternative model of critique:

> The immanent approach need not fear the objection that it is without a perspective, mollusklike and relativistic. Ideas that

> have confidence in their own objectivity have to surrender *va banque*, without mental reservations, to the object in which they immerse themselves, even if that object is another idea; this is the insurance premium they pay for not being a system.[6]

I know that I do not fulfil this ambition thoroughly or consistently enough; some material demanded the approach more than others. Still, I did not want *Dialectical Passions* to become just an overview of positions available in modern theories of art and culture; hopefully, from the very process of trying to "keep faith" with "the desideratum of immanent criticism" a level of understanding emerges that might not have done so otherwise.[7]

"Do you ever come except to criticize? Is nothing ever right for you on earth?" The Lord's question of Mephistopheles, in the prologue to Goethe's *Faust*, raises the charges of pessimism and complaint that are frequently encountered in discussions of negation.[8] They have been leveled at many of the writers on whom I concentrate, just as they are at "the spirit who always negates." Many of these authors share a suspicion of what may be called easy positions (including those that might be associated with the "counterintuitive" moves of theory), and they themselves tend to be regarded as "difficult." A recurrent theme in this study is the suspicion of figures of identity, immediacy, and plenitude, although the authors addressed differ considerably in how far they push their arguments, and in how they frame their contestation in theoretical and philosophical terms. Under the scrutiny of the most rigorously negative of writers, even simple pleasures can be rendered as the face of oppressive power; detecting the latent pure identity of nature, Adorno notes that "something frightening lurks in the song of birds precisely because it is not a song but obeys the spell in which it is enmeshed."[9] Goethe's Lord demonstrates considerably more patience with, and accommodation toward, the spirit of negation than do most of today's proponents of the affirmative outlook. An ideological continuum seems to join the psychotechnics of the neoliberal workplace and the pop psychology peddled by magazines ("Sisters! Every morning, look in the mirror and repeat five times: 'I love my thighs' "). It may be one thing to acknowledge the role of positive thought in a person's ability to survive serious illness, or to recognize its capacity to help extreme swimmers to survive the challenges of Arctic waters, but quite another to employ it as tool of corporate indoctrination—and still another to find it as a form of life-affirming aestheticism that seeks to ward off all critical engagement.[10] Often, the distinctive registers represented by these examples are simply collapsed together.

"Negation" is part of the routine language of art, and arguments about negativity are thoroughly embedded in accounts of culture and the debates on modernity and avant-gardism—and, by implication, those on postmodernity, or on the neo-avant-gardes and post-avant-gardes—frequently serving as vehicles for these discussions. The specific inflections of negation vary considerably. To take a few more or less random examples: the manifestos of Dadaists or Futurists (sometimes referred to as the "negative avant-gardes"); Gustav Metzger's "auto-destructive art"; statements such as "I would like my work to be a nonwork" (Eva Hesse); lists of denials deployed by artists from Ad Reinhardt to Vija Celmins; or Robert Smithson's fascination with entropic processes.[11] "The conditions for a passionate life existed, but I had to destroy them to be able to recuperate them" (Alighiero Boetti).[12] Or, the sculptor Carl Andre: "A man climbs a mountain because it is there. A man makes a work of art because it is not there."[13] Donald Judd's canonical definition of "specific objects" describes these works as being "neither painting nor sculpture"; the critic Barbara Rose described Minimal Art (or, as it is called, "ABC Art") as "a negative art of denial and renunciation"; the *via negativa* provided the focus for an Italian art exhibition in the midsixties.[14] For a period, Terry Atkinson makes "disaffirmation" the guiding principle of his work.[15] This brief inventory of negation's appearance in art could be readily expanded, and the idea of negation continues to animate approaches to contemporary art.[16] Of course, we would be mistaken to imagine that the remit of "negation"—a term that can be applied in and to any situation—is restricted to art or that it necessarily carries a radical charge (some of the examples cited here draw as much on Western artists' fascination with—and imaginative projection of—forms of Taoist or Buddhist mysticism, and Clark points to negation's aristocratic inheritance).[17] Nevertheless, the unusual prominence of negation in aesthetic discussions is notable, indicative perhaps of the long influence of Left-Hegelian and Nietzschean modes of thinking on the field, but also something of the particular nature, condition, and place of the art "object," of artistic activity, and of representation in modernity. The modalities of art (and what counts as art) may have changed substantially since Adorno drafted his book, but the opening lines of *Aesthetic Theory* still haunt contemporary art practice.

There is much that could be said here, but a couple of examples may serve to introduce at least some of the complex dimensions of this problem. Reality, the art critic Michael Fried notes, in a very brief and passing aside in his 1965 essay "Three American Painters," has come to outstrip and evade the representational powers of art.[18] Bertolt Brecht, echoing a comment made by the Marxist sociologist Fritz Sternberg, remarks that, since "reality has slipped into the functional," the photograph (the example given is one of a factory) could not capture the social relations of capital. Brecht's passage is famously cited by Walter Benjamin, and—in what becomes one of the founding theoretical moments for the strategy of aesthetic montage—they conclude that artifice and construction would be necessary to sidestep the inevitable flattening out of social complexity by the matter-of-fact image.[19] Preferring not to pursue his suggestive observation, Fried instead follows the classical high-modernist exploration of painting's "withdrawal from reality." There is a world of difference, of course, between arguing that art under modernism withdraws from reality (this is, after all, the heroic narrative that many modernists articulate) and saying that it is reality that eludes representation. The suggestion that reality slips into the "functional" and "abstract" harbors far more troubling implications, a profoundly social difficulty inscribing a situation in which realist ambitions for art seem irretrievable. Similarly, it is one thing to see painting or sculpture through this matrix (after all, the standard argument goes, photography could be left to get on with the routine task of picturing the world, releasing the higher arts to pursue their self-reflexive exploration of formal problems); it was quite another to challenge, as do Sternberg, Brecht, and Benjamin, the very ability of the photographic document to fulfill its designated role. The adequacy of these arguments—the one forged in Weimar Germany, the other in the ascendant hegemon of the postwar period—is not primarily what is at issue here; rather, the point is simply to recognize the extent to which "negation," in one of its guises, is implicated at every level of the problem outlined.

The ubiquity of negative expressions in art theory is matched by the seeming fluidity of reference. As Charles Harrison observes, negation is "promiscuous."[20] Taken in general, its meaning seems potentially infinite. Yet, in any one instance it is tied to the specific locus of its denial; the meaning of any "practice of negation" depends on what is being negated in the first place and, as Laura Mulvey acknowledges, cannot escape its defining ground.[21] This limited quality is frequently seen as a major drawback of such practices: they are unable to sustain a life of their own beyond their moment of resistance; if they succeed in their act of negating, they bring themselves down with what they denied. It is an argument employed not only in relation to cultural acts of negation but also to projects of social transformation.[22] Indeed, it is the very tension with such political implications that further animates the artistic and cultural discussions. Negation seems to be heavily suggestive by way of both its intellectual heritage—one thinks especially of Hegel, Marx, Bakunin, Nietzsche—and what Harrison refers to as its susceptibility to investment with "moral purpose."[23] A different way of approaching this would be to compare Max Horkheimer's description of Critical Theory as "formulating the negative" precisely in order to develop a "critical approach to existing society."[24] In *Negations* Herbert Marcuse describes "the sphere of negation, of contradiction to the established order, of protest, of dissociation, of criticism."[25] Its use is prominent in the writings of Henri Lefebvre, Jean-Paul Sartre; and, as the penultimate section of *The Society of the Spectacle* shows, negation is the language of choice for adherents of the Situationist International. In cultural and artistic debates, negation's weight is peculiarly complex: on the one hand, both highly specific and context bound, and, on the other, loose and plural; it is both meaning laden and meaning elusive. Its valence shifts from extremely particularized values, which require their specific determination to make sense, to a sort of reified metaconcept. Moreover, to speak of "practices of negation" in art is not the same as discussing "negative thought" and neither necessarily implies the more metaphysical, and sometimes theological, implications of deliberations on "negativity."[26] Nevertheless, there are some fairly direct connections between artistic practice and Critical Theory itself, with the philosophers—Benjamin being the clearest example—developing some of their most well-known terms from their encounter with avant-garde poetics. The symbiosis between Futurist artists and Russian Formalism's linguists and literary theorists finds later parallels in the work of Derrida and the authors associated with *Tel Quel*, which show a similar indebtedness to avant-gardist writing. As we will see in chapter 2, Tafuri argues not only that avant-gardist techniques relate mimetically to capital but that their aesthetic devices could even help propel forward its actual development.

Yet, very often, discussions exploit the slippages between these debates. Ranging across the metaphysical and scatological, idealist and material, the "moral charge" invested in discussions draws much of its force from this ambiguous potential, and the inevitable invocation of contentious philosophical-political legacies. This is probably why negation so frequently "leads to confusion," as T. J. Clark complains (in the light of seeing what his own deployment of the term unleashed).[27] Like the devil in Thomas Mann's reworking of the Faust myth, values metamorphose before one's eyes—and they do so not only in the sense of their constant multiplication but also by shifting qualitatively across the registers of general and particular. A local comment swiftly takes on the dimensions of a major social thesis; a minor act of aesthetic transgression magnifies into existential or world-historical significance. Often, general and particular categories are collapsed into a false identity. Nevertheless, one ignores this mutability and ladenness with social value at one's peril. So, while Harrison is correct to note that discussions of negation can elicit the worst sort of pietistic posturing—and while pursuing such discussions in the abstract risks seeming both portentous and ludicrous—it is equally futile to bracket out "moral purpose" in an effort to immunize oneself from these dangers. Considering negation within the parameters of formal logic does not seem to be helpful here; doing so would miss its characteristic life within cultural debates. By the same token, it would be inappropriate to seek a tighter definition, or, as Harold Rosenberg puts it ironically, to seek to improve one's success in pursuing speculative butterflies "by perfecting the instrument" of capture and weaving "a net with smaller holes."[28]

My study considers some of the responses in art history to the ideas of the New Left. Broadly, these may be characterized as attempts to draw out the social dimensions of cultural forms. The project is not new. Although not all its early representatives were socialists, determinedly social approaches to art were to become closely identified with "Marxist art history," around or through which the art historians of the New Left reformulated their projects. The "sociological approach," "sociology of art," or "social history of art," as Arnold Hauser variously referred to it, initially coalesced early in the twentieth century alongside efforts to shape an alternative communist culture.[29] Under determinate political conditions of international conflict, revolution, and civil war, a number of art historians transformed the neo-Hegelian and neo-Kantian approaches predominant in the field, taking them in a materialist direction. It would be a mistake, however, simply to equate this wave with orthodox Stalinism. Social-historical approaches to the study of art and culture emerged in a complex relation to the communist movement and in a critical relation to its economistic reductivism (and often also in reaction to its official aesthetic preferences). For example, while promulgating an explicit Marxist theory of art in a number of articles on art and architecture, Max Raphael kept his distance from the KPD.[30] In the United States, Meyer Schapiro—like Harold Rosenberg and Clement Greenberg—was formed in the intellectual orbit of Trotskyism. Those historians who developed radical approaches to the study of art belonged to the same generations as the major Left-wing intellectuals of the twentieth century: Raphael, Hauser, and Frederick Antal were born around 1890; Schapiro and Klingender were a decade or so younger. The history for some, however, was nonsynchronous. While important Marxist perspectives on art—including work by Raphael, Greenberg, and Schapiro—appeared through the nineteen-thirties, in the case of Hauser and Antal, their displacement as exiles—first from Hungary, later escaping National Socialism—interrupted their work and their major publications did not appear until after the war. Nevertheless, it is fair to say that the social-historical project typically crystallized during the thirties and Antal's own conception of these developments mentions writing almost entirely from the latter part of that decade.[31] By 1933–1934 Antal had settled in London, where, supported by Anthony Blunt, he worked occasionally for the Courtauld Institute; Hauser arrived in 1938. In a highly influential article on the intellectual makeup of English culture, Perry Anderson argued that not only did England lack a native Marxist current but that it had drawn its intellectuals from the White emigration; his example in art history is Ernst Gombrich.[32] The art-historical émigrés in Britain were in almost every case from Germany or the former Austria-Hungary (Gombrich, Antal, Hauser, Otto Pächt, Helen Rosenau, Gertrud Bing, Fritz Saxl, Edgar Wind, Rudolf Wittkower)—this was a Jewish emigration much more than it was a White one. Anderson also missed the political significance of that emigration. Gombrich was certainly to become the most well-known, but the approaches represented by the other émigrés—the cultural-historical approaches advanced by the Warburg school and the Red wing—were to be central to the later development of art history as a socially critical formation.[33]

Short lived it may have been, but the effects of the 1919 Hungarian revolution reverberated through the intellectual life of the twentieth century; both Hauser and Antal were shaped by it, becoming exiles after the counterrevolution. Earlier, both had participated in the Budapest *Sonntagskreis*, the famous Sunday Circle organized around Lukács and Béla Balázs, which attracted a number of important intellectuals. They were involved with the Free School of the Humanistic Sciences initiated by the group in 1917 and took up positions during the period of the Hungarian Soviet Republic (Antal as Chair of Budapest's Museum of Fine Arts, Hauser as a literary theory specialist at the Secondary-School Teacher-Training College, working alongside Karl Mannheim, another participant in the *Sonntagskreis*).[34] Lukács believed that of the Circle "only Antal had any Marxist leanings," and early relations with Hauser and Mannheim were often strained (Balázs recording how their commitment to serious action was doubted by others in the group).[35] Lukács continued to fault Hauser's philosophy for giving too much weight to historical contingency, a difference that resurfaced, amicably, in a radio discussion to celebrate the Hungarian translation of Hauser's *Sozialgeschichte der Kunst und Literatur* in 1969. By this time Hauser was perceived rather differently, many taking him to be a clear representative of Marxism. His *Social History of Art*, now in multiple translations and piled high in Maspero's Paris bookstore, had become a bestseller among radicals, even serving as an educational tool in Marxist history and social theory in countries where Left literature was officially proscribed; censors, it seems, did not wade through long histories of art.[36]

The social-historical approach to art and culture was also largely a response to the emergence of its polemical other. During the twenties, formalism was developing a more programmatic methodology and was being promoted with greater assertiveness. Explicit objections to interpretations of art due to "extrinsic" factors already featured in Heinrich Wölfflin's *Kunstgeschictliche Grundbegriffe* (*The Principles of Art History*) of 1915, synthesizing ideas outlined in his writings over the previous twenty-five years. In the following decade the young "radical formalists" of the "New" Vienna School—including figures such as Hans

Sedlmayr, Guido Kaschnitz von Weinberg, and Otto Pächt—sought to recover elements of Aloïs Riegl's project for a more rigorous scientific approach to formal questions. Presenting their work through the Berlin-based *Kritische Berichte* in the late twenties (a journal that Antal coedited with Bruno Fürst between 1926 and 1931), they founded their own journal *Kunstwissenschaftliche Forschungen* in 1933, a publication that drew the critical attention of Walter Benjamin.[37] Meanwhile, in the North American context, the modernist formalism that was to become so influential took shape. Greenberg was to develop his views in sympathy with the traditions of formal analysis, but Schapiro prepared Marxist critiques of the formalisms both of the New Vienna School and of Alfred H. Barr Jr., the first Director of New York's Museum of Modern Art.[38] The early social approaches to art primarily directed their criticism toward the method promoted by Wölfflin.[39] It was in response to Wölfflin's "art history without names"—art history as an anonymous process of immanent causation—that Hauser produced one of his most careful and sustained criticisms, the 1959 volume *The Philosophy of Art History*. Antal had studied with Wölfflin in Berlin (as had Raphael, and Benjamin had also attended his lectures) and subsequently undertook doctoral studies with Max Dvořák in Vienna, and it may be significant that Dvořák's work—and especially his attempts to relate artistic forms to wider patterns of culture—was itself drawing criticism from the representatives of the New Vienna School.[40] As a prominent version of formalism, Wölfflin's work clearly conformed to the sort of target that might be expected of mid-twentieth-century Marxists. Nevertheless, Hauser's reflections in *The Philosophy of Art History* certainly did not promote a simplistic model of superstructural reflection. (Antal's study of Florentine painting likewise demonstrated far more careful consideration of the relation between artifacts, ideology, publics, and economy than many imagine.[41]) Developing a complex and nuanced critique, Hauser's account was a probing—if unwieldy—analysis of the relation of Wölfflin's formalism to nineteenth-century historicism. Emphasizing the role of human agency, Hauser sought to avoid a method where "the particular is explained as a mere effulgence of a higher power, of a stronger purer light" or "which makes concrete history a reflection or a realization or an articulation of a universal metaphysical principle."[42]

To an extent, the occlusion of the first generation social historians of art within art history was moderated by the revival of critical Marxism in the late sixties and early seventies. Out of this renewed interest in social art history and Marxism, Raphael's *Proudhon, Marx, Picasso* was belatedly translated into English, while new work such as Nicos Hadjinicolaou's *Art History and Class Struggle* and his articles in *Kritische Berichte* engaged with Antal's writing.[43] There were a number of important developments: Clark's publications, and his part in initiating a postgraduate program specifically dedicated to the social history of art in 1976; the debates in (the new) *Kritische Berichte* and *Histoire et critiques des arts*; interventions from figures such as Karl Werckmeister, Kurt Forster, Jutta Held, Adrian Rifkin, and Hadjinicolaou; and the formation of the Marxist Caucus of the College Arts Association, to cite just a few examples. To these we could add the appearance of noteworthy non-Marxist contributions to social methodology by figures such as Michael Baxandall and Linda Nochlin; the development of feminist approaches to the subject by Nochlin, Rozika Parker and Griselda Pollock, Lisa Tickner, Norma Broude and Mary D. Garrard; the attention given by the film periodical *Screen*, especially in the seventies, to Russian Formalism, Brecht, Benjamin, and structuralism; and the emergence of new journals such as *October* (1976), *Block* (1979–1985), and *Oxford Art Journal* (founded 1978). Though these were to produce what often became competing intellectual tendencies, all contributed to the interrogation of the methods, purpose, and social function of art-historical work and were important in articulating dissatisfactions with traditional approaches.[44] By the mideighties, the broad range of new ways to study art became known under the umbrella of the "new art history."[45] In line with wider developments in the humanities, ambitious art history in this period became intensely methodological, but debates also bifurcated along the old fault line of history and philosophy, with radicals occupying both zones. By the end of the century, socially informed approaches to art and representation had become the predominant method in the discipline, with focused, local, scholarly studies forming the counterpart to an industry of more general introductory books (often structured according to a progression of art-historical methods) and a significant strand of self-consciously theoretical interventions (whose status lay somewhere between history and criticism). The basic assumption across all modes, and which mirrored the postmodern conception prevalent during the late twentieth century, was a suspicion—pursued and proselytized with different degrees of rigor—of all value claims, from "the aesthetic" and "the canon" to "art" per se. Indeed, once the "social" and "historical" became forces for relativism, these very notions were transformed. In the name of theoretical sophistication (and "antireductionism"), the complex relation between the general and particular frequently found itself collapsed into a narrowly "sociological" conception of art, if not dismissed outright.[46]

The explicitly antidialectical argument in art history was initiated by Ernst Gombrich—to those outside the field, perhaps still art history's most famous representative. Gombrich's broadside was occasioned by the publication of Hauser's *Social History of Art* in 1951. In his review for *The Art Bulletin* in 1953, Gombrich challenged both idealist and materialist forms of dialectical thought, drawing on a mode of argument that well suited the preferences of the Anglophone philosophy of his adopted homeland. The intellectual world of Critical Theory and Marxism that Hauser inhabited was theoretically incommensurable with the modern positivism and critical rationalism (of the Popperian variety) characteristic of Gombrich's approach. The "positivist dispute," which erupted as debates in German sociology in 1961, echoed the older opposition of Hegelian and Comtean social theory and had also been anticipated in the thirties by Horkheimer's critique of Vienna School philosophy.[47] These differences simmered through the review. Gombrich distrusted Hauser's efforts to assess art's "significance in the light of social theory," especially when that social theory was caught in "the intellectual mousetrap of 'dialectical materialism,'" and he objected to Hauser's conviction that "all factors, material and intellectual, economic and ideological, are bound up together in a state of indissoluble interdependence."[48] (In a reprise of an argument from art history's New Vienna School, Hauser's perceived failings were also attributed to Dvořák's influence.) The trap Gombrich identified was the ontology of "contradiction," which he thought led only to "theoretical paralysis."[49] As Gombrich understood it, Hegel "believed that reality was 'identical' with the process of reasoning"; thus, while an emphasis on contradiction might make sense in "the fantasy-world of Hegel's metaphysical system," materialists, such as Hauser, he argued, were less well served by an attachment to a mode of thinking that derives from seeing reality as the "thinking process of the Absolute."[50] The "Hegelian confusion" and "dialectical tangle" was not only "impenetrable" but also led to "logical absurdity."[51] A reliance on *Geistesgeschichte*, Gombrich thought, even bypassed the tasks of a social history of art, Hauser's "preoccupation with generalities" making him "careless of the significant detail."[52]

Some twenty years later Gombrich's complaints about the "lack of concreteness" and theory dependence echoed through art history's mainstream as the standard objection to any socially orientated approach.[53] While there were specific political and social issues at stake here, we should be aware that this concern with a lack of specificity has long reverberated throughout the history of art history: the art historians of the New Left, as well as Gombrich, found the approach of Hauser and Antal to be far too generalizing; yet this is exactly what Antal said of Dvořák, what Dvořák said of Wölfflin, and what even Wölfflin himself claimed to be criticizing.[54] The categories of German Idealism were perceived by the New Left to be one the major failings of the older generation. For Clark, as we will see, this critique remains allied to dialectical thinking, a commitment rejected by many others. In Gombrich's Cold-War-inspired and anti-Marxist dismissal of Hauser's "Hegelianism" we can discern preludes to the anti-Hegelianism of Althusserian Marxism which, as in the humanities more generally, became prominent within the new art history.

Published in the nineteen-sixties, Gilles Deleuze's reinterpretation of Nietzsche as a philosopher of affirmation rather than negation was a significant axis in the suspicion toward "the Hegelian dialectic," playing a key role in the emergence of poststructuralist thought.[55] This represents an interesting reversal since Nietzsche's appreciation among artists earlier in the century had been allied to what would later be perceived as their "negatory" practices. Deleuze's argument—just one enunciation of a new episode in critical and emancipatory thought—was subsequently abstracted from its specific context and assimilated into a general thesis that shaped the dominant consensus of much cultural theory of the late twentieth century.[56] Deleuze's foundational study was itself overdetermined by the French context of Left-existential philosophy, by the legacy of the Hegelian turn in the thirties, and by the central place of the PCF within intellectual life. The antidialectical tenor echoed within some sections of the New Left—from Louis Althusser to Lucio Colletti—where it was also identified with its own "immature" moments. In Althusser's case, the argument for a breach within Marx's thought—his "epistemological break"—combined structuralism with a scientistically orientated Marxism inherited from the Second and Third Internationals, and separated the "mature" Marx from his earlier flirtations with aspects of idealist (i.e., Hegelian) philosophy. While adding the novelty of structuralism to the mix (itself a response to Sartre and the legacy of Kojève's teachings), this touched on an old argument, returning us to the moment of Georg Lukács, Karl Korsch, and Antonio Gramsci in the nineteen-twenties, but also to Lenin's divorce from the mechanical materialism of the Second International in 1914, and to Marx himself, who, writing in his "mature" phase, explicitly reaffirmed his debt to Hegel.[57] In his early writings, Marx praised Hegel's *Phenomenology* for its focus on "the self-creation of man as a process" and, within this, on the role of labor.[58] As Marx put it in his "Theses on Feuerbach," idealism, not materialism, provided the "active" component of his philosophy and his conception of praxis. Contrasting with early materialism's contemplative character, these dynamic qualities from idealism were integrated with a politics of concrete sensuous activity. The ongoing tussle about how to weigh the dual legacies of idealism and materialism in Marxist thought, then, has formed another important facet of the antidialectical argument.

Yet even for those retaining sympathies for dialectics, and especially for its materialist versions, there has been no haven from its critique. Probably the most important aspect of the antidialectical argument emerged among the dialectical left itself, beginning with the contestation among Hegel's students—the "Young" or "Left-Hegelians"—in the eighteen-forties. Their emphasis on negation was inseparable from the political circumstances in which they found themselves, specifically the politics of German academia and its relationship to the Prussian state, which Hegel increasingly identified as the highest realization of spirit. (Marx later referred to this as the dialectic's "mystified form" which seeks "to transfigure and to glorify the existing state of things.")[59] How to evaluate Hegel's culpability has been keenly debated ever since, starting with the debate over his "exoteric" and "esoteric" views and progressing through to Derrida's discussion of the "general" and "restricted economy" of Hegel's writings.[60] Although mostly bourgeois republican in leanings, some Young Hegelians went on to found German socialism and communism. The Left-Hegelians attempted to salvage the dialectic's radical impulse. "In its true method," wrote Arnold Ruge, the dialectic "is criticism"; Marx later referred to its "critical and revolutionary essence" which negates the state of things and "regards every historically developed social form as in fluid movement."[61] Taking issue with Hegel's antiutopianism and his refusal to entertain anticipatory politics, they extended the dialectic of negation toward the future, turning it away from a passive interpretation of how the present emerged from the past, and instead taking it toward active criticism and then to action itself. Spurred perhaps by his admiration for Fourier and Saint-Simon, and by his meeting with Proudhon in the mid-eighteen-thirties, Cieszkowski mediated practice and philosophy with "praxis," making "will" its motive force (and thereby restoring a Fichtean aspect to the Hegelian dialectic, through the moral-practical idea of *Sollen*, or obligation, and the future-orientated will in action which opposes given reality).[62] Bruno Bauer's negative version of the dialectic has been seen as reintroducing radical dualism into Hegel, whose own thought seeks linkages—and not just antagonisms—between mind and world. As McLellan has noted, "the Young Hegelians gradually turned Hegel's dialectic, to which the concept of mediation was essential, into one which held all mediation to be anathema"—echoes of which we will encounter in chapter 2 among Italian thinkers of the nineteen-sixties and seventies.[63] Believing the failure of the French Revolution and of bourgeois republicanism to deliver "rational freedom" to be due to their faith in "constitutional mediation," Edgar Bauer's conclusion that "anarchy . . . is the beginning of all good things" could almost be a slogan for many of those on today's anticapitalist left whose faith in reason and the organizational forms of radical politics has also been shaken.[64] The degree to which negation is emphasized, to which mediation is associated with antifreedom, or to which affirmation is engaged—not to mention Marx's own inflection and critique of Left-Hegelianism—haunts the formulations and debates addressed in this book.

The classic discussion of negation occurs in the preface to the *Phenomenology*. Its context is an account of the relation between Substance (that which is) and Subject (that which thinks). Hegel felt that philosophy had separated these elements, yet "everything turns on grasping and expressing the true, not only as Substance, but equally as Subject."[65] Superficial thinking claims to immediately intuit God, which, Hegel thought, reduces divine life to "insipidity" and cheapens the Absolute by rendering it an abstract universal ("the abyss of vacuity"). The same thinking sees mediation as external and inessential, when for Hegel it is the process of mediation—"a becoming-other that has to be taken back"—that breaks from the illusion of the "first, immediately enunciated," which substitutes the beginning for the result. What superficial thought lacks, he contended, is "the seriousness, the suffering, the patience, and the labor of the negative." The labor of the negative (or the self-movement of form), he argued, actualizes or gives form to the Absolute through a process of movement and development or self-consummation. His emphasis is upon the movement of living Substance, its becoming-actual. Initially, this occurs through a procedure of "doubling," "bifurcation," self-alienation, or self-negation. But this "simple negativity"—the antithesis of "immediate simplicity"—is negated: self-alienation becomes self-return but a return not to the immediate but to a mediated unity. He goes on to make the same argument against "formalism" (empty abstractions and schema), against the sensuous immediacy privileged by "sound common sense" and empiricism, against "material thinking" and "picture thinking" which are unable, in his view, to extract themselves from "material stuff."[66] As he put it, one should wear neither "the robes of a high priest" nor "casual dress."[67] Where the ancients had forged the idea of the universal out of a challenge to natural consciousness, the task of the present is to break the inherited fixed forms now taken by the universal and to restore the fluidity to notions and ideas. (Adorno later described this as Hegel's essential battle with reified thought.[68]) Indeed, the strenuous efforts of the negative need to beware of false friends: the negative as merely negative; the negative as falsity or evil; the negative as external. Against the first, he noted that abstract immediate unity should be overturned by a determinate mediated unity, which gives determinate negation a "positive content."[69] The second error treats oppositions (such as good and evil) as fixed, ready-made, and mutually exclusive, to which positive and negative content is attributed. Without collapsing the two together, it is, he argued, important to recognize how the True has the negative "directly present."[70] Finally, because the negative presents a disparity between Subject and Substance, it is often mistakenly assumed that this gap is an externally imposed condition, whereas it should be understood as its own self-alienation. The ancients, he argued, had been correct in seeing "the void as the principle of motion . . . the moving principle as the negative," the movement of subject to object and object to object, but they did not recognize that "the negative is the self."[71] In one of the most famous passages, he wrote how the work of Understanding is an activity of dissolution. The "tremendous power of the negative" is revealed in the breakdown of elements and moments, their acquisition of some freedom from one another. The devastation resulting from the destruction of the immediate risks losing the idea to the unreal, "death";

but the evasion of this risk is to risk more, for only by "looking the negative in the face," or "tarrying with the negative," can truth be attained.[72] That so many debates return to Hegel, to his handling of negation, mediation, and to the controversies around reconciliation is indicative that, as one commentator has put it, "Hegel is the inaugural thinker of the contemporary world" and that "we read Hegel or we think him such as he has already been played out in thought."[73]

Reflecting on the history of Critical Theory and Western Marxism in the twentieth century, Perry Anderson has seen the increasing attention to aesthetic-philosophical concerns as signifying a retreat from politics.[74] Yet, we can read this same drift into considerations of art and aesthetics as resulting in an insistent social loading of aesthetic categories. In the same measure as it signaled a wider culture of political defeat, aesthetic discourse itself became (re)politicized, in a sense, returning discussions of the aesthetic to their modern beginnings. (Political aesthetics also responded—and can still be understood as responding—to the presence of aesthetic autonomy in both its ideological and institutional forms). The publication of the collection *Aesthetics and Politics* in 1977, translating debates between Benjamin, Adorno, Lukács, Bloch, and Brecht (and including Adorno's critique of Sartre and an introduction and afterword by Fredric Jameson), can be seen as indicative of this moment.[75] Anderson's comments captured something of the anxiety around aesthetics and the antagonism toward idealist subjects that were as prominent among radical art historians—especially on the German Left—as they were among those intellectuals outside of art who privileged politics. The rhetoric of negation still capitalizes on this powerful conjunction of social critique with aesthetic debate, although toward the end of the twentieth century, attention to aesthetic questions figures not so much a retreat from but the very presence of radical political aspirations, sometimes even their tentative return. So, for instance, we find the critical potential offered by German Romanticism and Idealism being addressed by Andrew Bowie and J. M. Bernstein.[76] Others—to take just a few examples—can be found arguing for renewed attention to the work of Adorno or seeking to reclaim the radical character of Benjamin's contribution.[77] Some look to the translations of Giorgio Agamben's writings, such as his analysis of the motif of art, to displace the longstanding emphasis on reception and spectatorship and reinstate the very activity of its producers.[78] We can also see how, by outlining the shape of our current "aesthetic regime," Jacques Rancière renews awareness of aesthetic autonomy's own fundamentally heteronomous birth.[79]

Aesthetic discussions, then, are increasingly sites of critical social reflection. Indeed, detectable among radical philosophers and cultural theorists is a prevailing sense of frustration with the barriers to critical thought. Alain Badiou argues that, for all its apparent historical innovativeness, the twentieth century provides us with no significant "new schema"—a declaration that, at first sight, may surprise those who have experienced or studied its political and theoretical rollercoaster.[80] Meanwhile, Rancière notes the limitations of the concept of "modernity"—a term around which so much discussion takes place—and which he criticizes not simply for its internal incoherence, but, more specifically, for the way that it serves to mask the history and specificity of our current "regime" of art.[81] Interestingly, the sense of blockage to which philosophers allude is mostly perceived and presented as a conceptual aporia. Fredric Jameson goes further still, arguing that "the modern" does not even ascend to the level of the concept but presents merely a terminological problem. Preferring to speak in a language of philosophical sublimation, many writers have seemed reticent to articulate the problem of conceptual hiatus in more direct political terms. For the most part—and perhaps more so than in its earlier manifestations— the contemporary discourse of "aesthetics and politics" adopts allegorical form. This situation has started to change. Jameson is increasingly prepared to speak in terms of social transformation, pointing out that until we abandon "the conceptual field governed by the word 'modern'" we will not be able to imagine, let alone theorize, "systematic transformations."[82] Jameson's sights are specifically turned on the political discourses of modernization associated with, for example, the proselytizers for the "third way," but it is equally relevant to understand his argument as challenging an intellectual radicalism focused upon the problem of modernity. "Ontologies of the present," he concludes, turning inside out the anticipated vectors of temporalization, "demand archaeologies of the future, not forecasts of the past."[83] More unashamedly political and optimistic assessments have emerged from post-*operaist* and post-Autonomist writers such as Negri and Hardt, perhaps because their arguments—as typified by *Empire*—essentially provide a Left gloss on the more familiar postmodern theory, thereby offering a recovery of post-1968 radical discourse and a responsiveness to the current moment while not actually shaking too many of the familiar theoretical touchstones of recent years. Yet, the decisive moment of *Empire* seems to have been to relegitimate openly political discussions within contemporary Critical Theory and to rupture the prevailing despondency. Despite the significant limitations of their argument—and there are many—Hardt and Negri's intervention gives renewed emphasis to questions of social process, a theme long absent from most discussions and especially those more pessimistic models drawn from the Frankfurt School. With *Empire*'s appearance, the mood of debate has been changing markedly, driven as much by the critical responses galvanized to challenge its core theses as from those immediately enthused by its claims. The trajectory taken by the subsequent volumes has clearly been driven by the dialogue with these Left critics, propelling discussion yet further from the residual cultural-theoretical consensus.[84]

We can speculate on what has brought these approaches to the fore and what has emboldened a growing number of authors to disturb "the heavy slumber of orthodoxies":[85] the resurgence of radical energies in response to the effects of the collapse of "historical communism"; the expansion of the neoliberal agenda; and the accompanying crowing of capital's loyal commentators; not to mention the systemic crisis that became starkly visible in 2008. Debates on globalization, antiglobalization, and alterglobalization; growing concerns over the climate and environment; and opposition to the invasions of Afghanistan and Iraq raised issues that could not be readily accommodated under established rubrics. While these militant voices are yet to displace the "disabling eclipse of social imagination" that is "manifested in the almost universal assumption of a continuing capitalist future," a new constellation of critical thought is certainly emerging.[86] It is not just the dominant historical narrative of neoliberalism that is being disrupted; the internal intellectual hegemony within emancipatory discourse itself is again being contested and reconstituted.

With the protests against the WTO summit in Seattle in 1999, many argue, the postmodern consensus was brought to a definitive end.[87] T. J. Clark's *Farewell to an Idea*, published the same year, still seems caught on the threshold of this transformation.[88] The author takes leave of "an idea"—the project of modernism and socialism—but in a way that seems decisively framed by the cultural-political consensus that he had long resisted, his conclusion rehearsing another "adieux aux proletariat." One wonders how different the tenor of his introduction and conclusion might have been even just one year on. Certainly, the tone of *Afflicted Powers*—in which Clark was involved as part of the Retort collective—is far more trenchant. Initially prepared in the form of a broadsheet, *Neither Their War Nor Their Peace*, for the antiwar protests in February 2003, *Afflicted Powers* adopts a polemical voice. It is framed by Satan's—and debatably also John Milton's—reflections, in the first book of *Paradise Lost*, on how to recover from defeat and dispersal.[89] *Afflicted Powers* posits a period that sees a proliferation of ruptures emerging within "the spectacle"—even the potential for its internal collapse. Registering a number of profound failures at the level of the spectacle, the authors point to the fractures opening in the seemingly impenetrable armor of "military neoliberalism." The claim that we entered a "new 1914" along with the invocation of Rosa Luxemburg's *Junius Pamphlet* and the symbolism (afforded by the warmongers' own timescale) of early springtime demonstrations serve to dramatize new resistive potential, even as the authors engage critically with prominent arguments *within* the contemporary anticapitalist movement, and even as they caution against the voluntaristic optimism offered by Negri and Hardt.[90] No one anticipated, however, that within just a few more years mainstream commentators would be comparing our everyday experience with 1929.

It is not as if, in some absolute sense, everything just changed overnight. Some writers, for instance, have identified the uprisings in Chiapas at the beginning of 1994 as the turning point; others have alluded to French protests over the winter of 1995–96 as a moment of fragile and "frail rebirth."[91] The shift can be detected in art too. In 1997, Catherine David oversaw the curating of Kassel's tenth quinquennial mega-exhibition, Documenta X.[92] Drawing on radical political or socially orientated projects from across the twentieth century, she presented art that emphasized the political, ideological, and social geographies of the modern and contemporary world. Okwui Enwezor substantially extended this politicized lineage at the next Documenta in 2002, foregrounding, in addition, issues of globalization, creolité, truth, and democracy. But a great deal had altered in the five years dividing these two events. David's Documenta X, with its unapologetic centering on the spirit of the midsixties (which she cycled backward to the close of the Second World War and postwar reconstruction and forward to the end of the twentieth century), was widely criticized at the time for being a form of outdated Left-wing nostalgia (and the commodification of it to boot); just a few years later, her curatorial intervention was seen as being remarkably prescient. Buoyed by a growing body of art taking critical and political orientations, themes addressing capital accumulation, privatization and new enclosures, global migrations and population movements, national borders, the repression of historical memories and the imaginary of social transformation were pursued in prominent exhibitions elsewhere.[93] Such changes in the emancipatory discourses of social and cultural theory should not be underestimated, and we find that categories deemed irremediably passé for much of the late twentieth century, such as totality, universalism, and truth, have again come back into serious critical focus.

Intervening in radical debates on contemporary art theory, *Dialectical Passions* offers a fresh perspective on, and challenge to, the specific closures that developed in the wake of the New Left and through what could be called "the long nineteen-eighties." Here I have in mind the reshaping of the postwar order and, especially, the rise of political and economic neoliberalism that centered on—but was not confined to—that decade, and I am particularly interested in how radical reflections on art and culture, and the political philosophies associated with them, became reframed by this situation. "Postmodernism" is probably the phenomenon most closely associated with this period, but although the modern-postmodern debates are an inevitable point of reference for many of the debates I discuss, it is not postmodernity per se that interests me. (I have always been skeptical of the claims made for it, doubting the characterization of the various artificial targets that were set in its sights.) Not so long ago, postmodernity was presented to us as our total horizon; not so long before that, no one even knew it was coming; now it seems remaindered as a concept, let alone as a descriptor that might adequately delineate a particular period or cultural *Weltanschauung*. At first sight, Benjamin's project to reclaim the past for the present, to deploy the traces of the long-gone to erupt into now-time, seems equally redundant; instead, we seem to face the prospect of the just-gone receding rapidly into the far-past. However, this prospect makes Benjamin's point all the more forceful and we now need such interventions into our own immediate yesterdays. *Dialectical Passions* attends to the erasures and recoveries of critical memory that occurred in the passages between these "not so long agos." It reconsiders the production, and questions the hypostatization, of the account of the relation of modernity and capitalism to the critical assessment of aesthetic and cultural forms, and addresses how we might release some of the blockages that have come to beset radical thought and its horizons, so that we might better understand our own situation and revive our critical agency.

Chapter 1

T. J. Clark and the Pain of the Unattainable Beyond

> Complicated, full of interruptions and contradictions the way may be that leads from certain social conditions to the creation of spiritual values, as for instance from Dutch middle-class capitalism to the works of Rembrandt; still, in the end one has to decide whether or not such conditions are relevant. One can put off the decision, conceal one's position, talk of dualism and dialectic, reciprocity and mutual dependence of spirit and matter; but after all one is either an idealist or a realist, and has to face the question of whether genius falls from heaven or fashions itself here on earth.
>
> —Arnold Hauser[1]

Part I

Since the publication in 1973 of his first two books, *Image of the People* and *The Absolute Bourgeois*, T. J. Clark has been identified as among the most significant art historians of the postwar period. As testified to by numerous courses and readers in art history, these two works—the book on Courbet especially—are seen as a turning point in the discipline, initiating the second wave of "the social history of art," and providing a point of departure for the more broadly conceived "new art history."[2] The method advocated by Clark in the opening chapter of *Image of the People*—a form of social-historical work that pays close attention to art's field of discussion, looking for ruptures and elisions within art criticism—soon acquired the status almost of a disciplinary reflex, especially for the art history of the modern period. This might be reason enough to consider his work, but it is Clark's place as a thinker of the Left that will be of most importance here. This chapter does not attempt to summarize the content of Clark's accounts of artists such as Poussin, David, Courbet, Manet, Pisarro, Malevich, or Hofmann, but rather explores a deeper logic in his project: the central role of mediation in the social history of art, and that of the equally central, though increasingly displaced, idea of negation. The problem of mediation is a continuing question for an understanding of Clark's project, and it will be explored primarily, although not exclusively, through his writing on Jackson Pollock, first appearing as an essay, and subsequently reworked for *Farewell to an Idea*.[3] The account of Pollock's abstract painting is particularly instructive because it displays Clark's method *in extremis*, stripped of the more obvious historical references, crutches, or clues supplied by resemblance or "subject matter." When the work of art consists of nothing but drips and splatters of paint, Clark's desire to discover the "network of real, complex relations" that embeds the painting's sociality and historicity is put under a distinctive set of strains. When focusing on the problem of mediation, changes in Clark's methodological points of reference—whether Macherey, Freud, Bakhtin, or de Man (and leaving aside the ever-present figures of Hegel, Marx, Lukács, Benjamin, and Debord)—signifies a set of persistent concerns rather than an abrupt change of course or, as some have assumed, an abandonment of a social-historical project.

In his writing, Clark does not systematically or cohesively focus his theoretical inquiry on the theme of mediation. There is intention at work here, for sure—mediation for Clark is not a repressed or unconscious drive, but neither is it a direct "object of study"; rather, mediation is something that just seems to bother him and it emerges in his writing like a perpetual gnawing. Clark is lured by his subjects, often down labyrinthine paths; indeed, it is through his material that his dialectics seems to come alive. In this chapter, a series of aporia in Clark's work will be unfolded. They represent neither petty inconsistencies nor casually introduced contradictions; rather, they result from the high stakes of the project and its arguments, emerging as difficulties produced by thought and its relation to its object at a particular historical conjuncture.

It was the mideighties before Clark followed the success of the early books with *The Painting of Modern Life*.[4] This study, focused on Manet and Impressionism, elaborated themes from Meyer Schapiro's pioneering essay of 1937, "The Nature of Abstract Art": the structures of class, social space, and leisure in late-nineteenth-century Paris.[5] In the intervening years the critical reorientation in which he had been a key player had further shifted its terms of reference; Clark's unapologetic attention to the canon of painting, as well as his essentially Marxist project, sat uneasily with the seemingly more dynamic methodologies of poststructural and feminist analyses and with the turn to visual cultures. Among erstwhile radical supporters, as well as the already disgruntled conservatives, the book was not well received. Many were exasperated with Clark's approach: the sweeping judgments of value; an exclusive attention to the privileged objects of art, which relegates all other visual forms to a supporting role; his preoccupation with individual artists, which seems too close to a concept of "genius"; and a focus on canonical male artists. Nor did his type of questions concur with the protocols of the new critical art history, not least, his presentation of an apparent partition between art and society or between text and context—that is, the very divisions to be addressed by the act of mediation. Nevertheless, while his writing has been subject to a wide range of criticisms, there remain substantive questions to be probed that are immanent to his approach and that turn precisely on this problem of mediation.[6]

With its large themes and its provocative title, *Farewell to an Idea* firmly propelled Clark's work beyond the borders of his discipline. Confronting the questions of modernism in art—its relation to both capitalism and socialism and the extent to which social values are carried by painterly abstraction—it is, surprisingly, among the very few full-fledged reckonings with modernity available in art history. Yet, with its self-conscious episodic structure, its account of modern art since David essentially comes to a halt with Hans Hofmann. It is notable that Clark fails to tackle anything from the last half of the twentieth century. (His next book turned to Poussin.[7]) Indeed, there is very little in his writing that helps his readers discover what he thinks about the art that, for many of them, covers their entire lifetime. This is astonishing given that *Farewell*, in particular, is both presented and taken as a claim on the present, as an assessment of the current state of affairs, and as a judgment—for both politics and art—on the openings or closures of historical possibility.[8] It is as though there is a hiatus between the political stakes in Clark's argument and his critical engagements with the visual-material forms of contemporary expression. Clues can be found in a pamphlet cowritten by Clark as part of the English Section of the Situationist International in 1967: the art of the time is certainly found wanting; Happenings and Pop are seen as just forms of decadent nihilism and degenerate avant-gardism.[9] It still makes for a good read, even if today's sensibilities might balk at the authors' ease with these adjectives. And it is an ease that belies the deeper links with the values of the old Left and *Kulturkritik*,[10] a rendition of Lukácsian mores before the time when Adorno, Bloch, and Benjamin—and, it would seem, Trotsky and Gramsci—had complicated matters by finding critical value in modern art. Adorno et al. had in mind the art of Expressionism, Futurism, Constructivism, Surrealism; the problem facing the English Section, however, was the neo-avant-garde art of the sixties. Judgments about art in this period took place within a new context—not just a world of mass media, but within a mood shaped somewhere between populist incomprehension, the defensiveness of the major American critics (Greenberg, Rosenberg), the aesthetic preferences traditionally advanced by the Left, and, crucially, the growing sense of cultural commodification. Clark has commented ruefully—although unapologetically—on the aesthetic judgments he made back then:

> I realize—this realization has come upon me lately—that even in the period of my life when I did feel a vital connection with the art going on around me I got the art wrong. My militant hostility to Andy Warhol, for example, has not stood the test of time. I wish in retrospect that I had been able to recognize the strangeness and interest of Don Judd's sculpture. And so on. It is not that I fail to remember why I made these mistakes; and part of me persists in seeing them as necessary, or at least productive—for myself and others back then in the heat of the moment. It was the sixties. The question of art was posed for a while in extreme form, as an all or nothing choice. I know why I opted for the "all." And of course those of us who did so knew very well at the time that opting for the all—the all of the end of art, and art's realization in revolutionary practice—might end up as opting for nothing. We were not fools; we knew the cards of culture were stacked massively against us.[11]

This sense of intellectual and aesthetic production at its most tensile, hinting at "the end of art," is typically Clarkian. As we will see, the sense of historical straining displayed in this passage—the excesses, the testings, the forcings, the exacerbations and pressures in what he has called modernism's "wind tunnel" or "crucible"—pertains as much to the practices and problems of social transformation as it does to art.[12]

Clark's early publications are already marked by that very particular haunting voice that characterizes his writing: the lilting rhythm of self-unraveling ruminations, elliptical thoughts, and his extensive use of the first person, a mode of address that enthralls or irritates his readers in equal measure. Assuming those readers have enough grasp on Marx's *Eighteenth Brumaire* and the opening of historical and aesthetic possibilities in mid-nineteenth-century France, they can follow Courbet's negotiations of the upheavals of 1848 and his attempt to hone an art that might be equal to historical events. It is now impossible to look at *The Burial at Ornans* without thinking of Proudhon. The point, however, is not so much that the author of *The Poverty of Philosophy* influenced Courbet, but that Clark makes Proudhon's voice (and those of other individuals and critics) speak through the artist and his work. Clark, as he himself describes in the opening chapter, seeks to read through the criticism of the time with an eye to those condensations and displacements, shifts and jumps, which might allow a painting to collapse into, and then reemerge from, the complex discursive web that had been spun. If "History" is "Art's" negation (or even vice versa), then Clark hopes—by means of a Freudian hermeneutic of symptoms and textual metaphors combined with a radical Hegelianism—to arrive at the negation of the negation, to find the points where paintings concentrate their historical moment.

In his explanation of the value form, Marx makes commodities speak, allowing them their fetishistic rights as a means to strip them of their claims over human life. Clark turns this rhetorical device into method: ambiguous elements from a painting by Pollock or a Picasso converse; ducks talk with rabbits; awkwardness has a discussion with aesthetic delight; a brushstroke speaks to a social contradiction. At certain moments, the author resorts to yet more forced ventriloquism, pushing words into muteness, be it the silence within a critic's account or the monosyllabic, adolescent recalcitrance of an abstract painting. Over time this strained ventriloquy seems to have intensified. With Courbet and with David, say, the throwing of voices sits reasonably comfortably, even organically, within the canvases' depicted figures; it is clearly not so easy with a Pollock, Hofmann, or El Lissitzsky. More recent art may pose a greater challenge: even if many artists now reject the type of abstract art characteristic of the mid-twentieth century this has only been the exit from "abstraction" in its most obvious sense. The issue at stake here may well be social rather than one simply defined by some proscription on figuration. As we will see, Manfredo Tafuri also has much to say about El Lissitzsky's relation to the "disenchanted sign," and in *Farewell* Clark grapples extensively with something like the same problem, in a way that seems to be constitutive both for the artist and writer.

Clark's voice also displays the distinctive sensibility of a militant perspective on the world. We will, following Clark's own warning, have to be cautious in our use of the word "Marxist." His politics allow him to accept Marx and much that might be understood as Marxism, but given that he often identifies the latter term exclusively with the Third International and its various legacies, he is reluctant to situate himself within that frame. His critical sympathies for certain forms of anarchism, but more especially for noncentralized revolutionary or direct democracy, put him at odds with many others on the Left.[13] Over the years, old council communists, especially those once associated with the Situationist International, are likely to have experienced a rough ride from Stalinists, Maoists, Trotskyists, and their multiple postmodern successors and detractors. Clark's writing has a somewhat embattled sensibility and he anticipates critique—whether art-historical or political—in advance, embedding the imagined dialogue in his prose. This, after all, is a man who when picking up an ancient copy of *New Left Review* is almost sick on the rush of *memoire involontaire*.[14] And while one might empathize with a fair range of "infantile disorders" and their motivations—and feel pain over the tragedy at the mouth of the Neva—Clark's position can, nevertheless, have just too much of the "beautiful soul" about it. (No doubt, he is used to this criticism.)

In his opening chapter of *Image of the People*, Clark characterizes his project as "the social history of art." He invokes the work of Arnold Hauser, whose *The Social History of Art*, a massive study running to nearly one thousand pages and spanning from Paleolithic cave painting to "the Film Age," appeared in 1951.[15] As Clark claims in 1981, however, his appropriation of Hauser's title is meant ironically. It has rarely been taken so—ever the problem facing irony. (Is not the title of *Farewell* yet another of these buried ironies, only now posed in still more doubtful tone?) Some of the reasons why Clark uses the "social history of art" in this way should be clear from the previous discussion. The preferences of the first wave of social historians of art; their sense of art's history; their use of categories (such as realism, reflection, and the popular); their being in the orbit of the official Communist Party; their conceptualization of "base" and "superstructure," and of progressive art attached to rising classes and degenerating modes to declining ones, all signaled values that were undergoing significant reassessment by the New Left. Clark is a key participant in this transformation, but, as we have seen, he also refuses certain aspects of the conceptual shifts that became favored through the nineteen-seventies and eighties. He continues to mark his distance from the "academic Left" and from the "sententious self-righteousness" of "comfortable first-world professionals."[16] Noting, of the period 1848–1851, that "Bourgeois society is efficient at making all art its own," Clark seems to connect this statement to the long political fallout from 1968, expressing a self-consciousness of his own part in the academic "recuperation."[17] This conjunction of politics, a history of praxis, and the question of the historian's own role therein, is one factor that makes this body of work particularly compelling. An academic displacement it may be, but Clark's interest in mediation teems with the problems of what might be (contemporary) social transformation.

On "Practices of Negation"

> This phrase [practices of negation] . . . seems to have given rise to some misunderstanding, among those who approved of it as much as those who thought it a dreadful slur.[18]

> I do not like that formula any longer. I think it wrong to opt for either "negative" or "positive," or "beautiful" or "ugly," as descriptions of modernism in characteristic mood. The point is that modernism was always on the lookout for the moment, or practice, to which both descriptions apply.[19]

As these two statements—from 1984 and 2002—indicate, Clark has long been anxious over his use of the phrase "practices of negation." The history of Clark's relation to negation is worth attending to and takes us in two directions. "Practices of negation" nearly always crops up where Clark comes closest to outlining his theory of modernism, and continues to do so despite—even because of—his disavowals. The discussion also takes us further into the chapter's substantive topic: the processes of mediation in Clark's methodology. There is a protracted history to Clark's concern with negation. In 1969 he notes how the art history of the time tended to shrink from the works of Courbet, Rimbaud, and Lautréamont, and, in particular, from their "gestures of renunciation."[20] This point can be seen to continue an argument of the English Section. "Negation" is not as prominent in this essay as it is in Debord's *Society of the Spectacle*, but it is there, and in fairly unequivocal terms:

> The project of art—for Blake, for Nietzsche—became the transvaluation of all values and the destruction of all that prevents it. Art became negation: in Goya, in Beethoven, or in Géricault one can see the change from celebrant to subversive within the space of a lifetime.[21]

Negation here indicates both the crisis of value and a stance against the establishment. The position attacks the "illusory revolts" of contemporary avant-gardism, which, the authors argue, come in the forms of "reformism and nihilism" (themes that will return in the Tafuri chapter). As suggested by the pamphlet's chiasmatic title—*The Revolution of Modern Art and the Modern Art of Revolution*—there is a sense that aesthetic negations must be fulfilled by practical social revolution. In a phrase reverberating with the echoes of Surrealists from André Breton to Suzanne Césaire, and, of course, with the words of Debord, they note that "Life and revolution will be invented together or not at all."[22] The play on Surrealism's rhetoric is indicative of the SI's love-hate relation with politicized avant-gardism and its critical opposition to the mainstream of the Left. The opposition to the evil twins—Nihilism and Reformism—continues to be important in understanding Clark's position; at this point, Nietzscheanism would not be tolerated (an outlook he later modified).

The issue of negation is also developed in Clark's 1980 article in *Screen*, "Preliminaries to a Possible Treatment of 'Olympia' in 1865."[23] Manet's painting, Clark suggests:

> erodes the *terms* in which the normal recognitions are enacted, but it leaves the structure itself intact. . . . To escape that structure what would be needed would be, exactly, another set of terms—

> terms which would be discovered, doubtless, in the act of unsettling the old codes and conventions, but which would have themselves to be *settled*, consistent, forming a finished sentence.[24]

In this essay he distinguishes his own claim that Manet's *Olympia* "refuses to signify" from the "dis-identificatory practices" explored by many of the authors in *Screen*.[25] This is one of the arenas in which practices of negation open onto a wider debate. *Screen* was then an important forum in the revival of theoretical interest in Benjamin, Brecht, Russian Formalism, LEF, and other avant-gardist practices of the twenties. The interest was not simply historical; these avant-gardes of the interwar years were understood as having relevance for contemporary artists and filmmakers. The work of Jean-Luc Godard, and particularly his collaborations with Jean-Pierre Gorin in the "Dziga Vertov Group," was pivotal. Disruption of established narrative structures and the interruption to the lures of spectatorial pleasure were taken as a political strategy of cultural resistance to the dominant codes of mainstream cinema. In the wider context of political interventions that included "countercinema" and "guerrilla cinema," the approach promoted in *Screen* stood out for its explicit avant-gardist rupturing of filmic "form" and its attempt to negate ideological circuits set in play by the interaction of form, narrative identification, and spectatorial address. Aspects of the debate were picked up by a number of artists, particularly among those working with the photographic medium.[26] Through his discussion of *Olympia*, Clark distances himself from disidentification:

> any critique of the established, dominant systems of meaning will degenerate into a mere refusal to signify unless it seeks to found its meanings—discover its contrary meaning—not in some magic re-presentation, on the other side of negation and refusal, but in signs which are already present, fighting for room—meanings rooted in actual forms of life; repressed meanings, the meanings of the dominated.[27]

In the context of *Screen*, Clark's argument was encountered as a rerun of Russian debates from the late twenties or those among German Marxists of the thirties concerning the relation between modernism, realism, and Marxism. Peter Wollen was not entirely mistaken in detecting traces of Lukács, but the debate's terms had shifted slightly. Clark seems to challenge *Screen*'s avant-gardism with a realism—not, however, with one of recognizable types and figures but with a realism that is primarily concerned with the reconstruction and reoriention of meanings. But to say this is a disagreement between avant-gardism and realism is to overcode and oversimplify it: *Screen*'s avant-gardism is pulled toward realism, just as Clark's apparent realism is drawn to its opposite; with both sides being post-Brechtian and post-Lukácsian, this is a dispute over different radical realisms. The issue at stake might better be seen as a conflict between two senses of negation: internal and external or determinate and indeterminate.[28] Yet even this will not suffice. Although it might seem that Clark plies the route of an internal, determinate negation—specific resistances to specific entities and meanings—it turns out to be, as I noted earlier, a close collaborator of Wollen's, Laura Mulvey, who concludes some years later that the model of negation adopted by *Screen* was itself stymied by its dependency on the structures of identification and belief that it sought to counter. Disidentification also involves breaking from—and working against—something, perhaps too determinately.[29] In the end, the debate was left in an unsatisfactory standoff, the cognitive tensions it contained remaining unproductive, or worse, propelled by many readers back into a stark opposition.

A second arena for the discussion of practices of negation occurred soon after. Where *Olympia* and Godard, and the questions of realism and avant-gardism, had framed the *Screen* debate, this second discussion revolved around a critique of Greenberg's art criticism. Clark's engagement with Greenberg—which was first titled, with a degree of ironic provocation, "More of the Differences Between Comrade Greenberg and Ourselves"—was delivered at the conference "Modernism and Modernity" held in Vancouver in 1981. The mood of the Vancouver discussion was motivated by a deep dissatisfaction on the Left with the type of emerging postmodern critique that dismissed modernism *tout court* and the conference set as its task the recovery of a more complex and contradictory picture of modernism. The aim was to challenge Greenberg's restricted model of modernism, which in turn was often assumed by its postmodern rejecters. In both respects, negation became the operative category for the occasion: it was deployed to challenge the high modernist account, and, in a riposte to the new celebrants of postmodernism, negation was also deemed, as Serge Guilbaut put it, to be the "critical core" of modernism since the nineteenth century, "a coefficient of resistance to the prevailing system."[30] At a symposium organized in Chicago later the same year by *Critical Inquiry*, Clark presented his paper again—now with the prosaic title "Clement Greenberg's Theory of Art." A response from Michael Fried, published in the journal in 1982, was followed in turn by a short rejoinder from Clark.[31] By the mideighties these papers had been anthologized and widely disseminated.

This exchange is still widely perceived as the hub of Clark's engagements with practices of negation. Fried argues that Clark's critique of, and alternative to, Greenberg's theory of art is especially ill equipped to deal with abstraction, invoking, amongst others, Jackson Pollock's painting *Lavender Mist* (see figure 1.1). In his reply, Clark outlines how his critique of Greenberg's modernism is essential to understand what

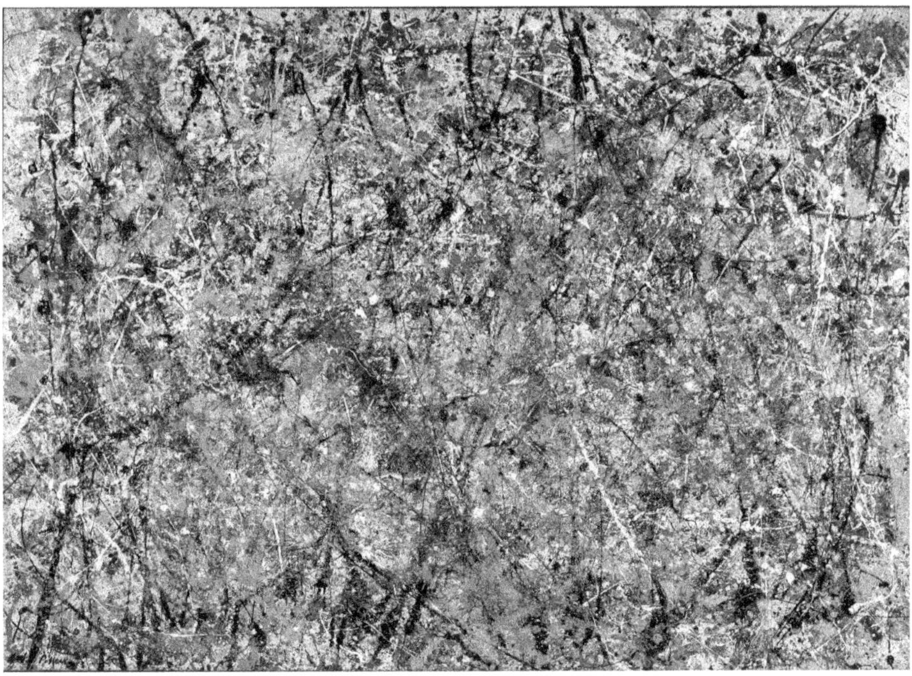

1.1 Jackson Pollock (1912–1956), *Lavender Mist: Number 1*, 1950, oil, enamel, and aluminum paint on canvas, 221 × 299.7 cm, National Gallery of Art, Washington DC, USA.
Photo: The Bridgeman Art Library © The Pollock-Krasner Foundation ARS, NY, and DACS, London 2009.

he calls the "delectable impasse" reached by Pollock's paintings by the end of the nineteen-forties. Clark introduces the phrase "practices of negation" in this context, in an effort "to recast [Greenberg's] sketch of modernism's formal logic."[32] Practices of negation, he argues, rework that logic to reopen it onto "history" and "the social." As Clark characterizes it, the classic modernist account of art (associated with Barr and Greenberg, and later with Fried), describes a sequence of "triumphant openings on to fullness and positivity."[33] Bringing the concept of negation to the fore, Clark couches Greenberg's account less as "purification" and more as difficulty and desperation. In doing so, Clark also opens up the term "modernism" from Greenberg's notoriously limited range of aesthetic preferences. "All the way to the Black Square"—to use Clark's (again slightly ironic) formulation—sums up a painful, albeit productive, process of elimination and renunciation. Malevich's black square involved the literal and physical obliteration of one of his own Suprematist paintings, but Malevich hardly counts among Greenberg's greatest hits. The ideal form of the square, then, embodied the utopia of the one of the very avant-gardes ignored by Greenberg; and it playfully terminates at an alternative—a more historically loaded—telos.[34] But the phrase "practices of negation" caused some confusion among Clark's readers, and with the republication of the essay in 1984, Clark attempts to clarify his intended meaning with an extensive author's note:

> By "practices of negation" I meant some form of decisive innovation, in method or materials or imagery, whereby a previously established set of skills or frame of reference—skills and references which up till then had been taken as essential to art making of any seriousness—are deliberately avoided or travestied, in such a way as to imply that only *by* such incompetence or obscurity will genuine picturing get done.[35]

Clark appears to have been concerned that his references to practices of negation were being taken by a number of his readers as a Left political program. But, whatever he claims he meant to mean, the identity of negation in his writings does not resolve itself so easily; Clark had already unleashed an unruly pack of associations.

Initially, on revisiting the first full articulation of practices of negation, everything seems straightforward and to confirm Clark's supplementary note: such practices are primarily enactments on, or of, an artist's medium. Modern artists' repeated rejection of accepted subjects, procedures, and competencies allows Clark to cast negation primarily as a process "of the medium," as specific and internal to artistic practice. This interpretation is affirmed in the 1984 author's note, and in statements such as: "the medium has appeared most characteristically as the site of negation and estrangement" and "practices of negation . . . seem to me the very form of the practices of purity (the recognitions and enactments of medium) which Greenberg extols."[36] Clark presents these practices of negation, then, neither as a procedure of controlled self-cleansing nor as a plenitudinous dialectic of artistic discovery of medium, but as "a whole strategy of release, exacerbation, emptying, and self-splitting."[37] He states his admiration for these struggles of "casting off" and for "the ruthlessness of negation":

> [Modernist] practice . . . is extraordinary and desperate: it presents itself as a work of interminable and absolute decomposition, a work which is always pushing "medium" to its limits—to its ending—to the point where it breaks or evaporates or turns back into mere unworked material. That is the form in which medium is retrieved or reinvented: the fact of Art, in modernism, *is* the fact of negation.[38]

In contrast to Greenberg's story of modernist painting (or at least its familiar interpretation), the progressive focus on medium is seen here as a deeply troubled and difficult process.

At the same time, however, Clark also describes negation in a very different way, as something whose force essentially derives from issues lying beyond medium. In this sense, paintings—and Clark is always interested in two-dimensional art—become articulations of values whose negative value resides at the social level. Furthermore, Clark discusses this social aspect of negation in two distinct ways. In the first of these accounts, he treats negation as an active social or political engagement by artists, what we might call, the values of social resistance:

> [T]he avant-garde . . . has regularly and rightly seen an advantage for art in the particular conditions of "ideological confusion and violence" under capital; it wished to take part in the general

> untidy work of negation and has seen no necessary contradiction (rather the contrary) between doing so and coming to terms once again with its "medium."[39]

Negation here means playing for the stakes of the modern world, the desire to participate in changing the social conditions of life, and goes alongside the radical exploration of form. It is worth comparing this sense of negation with Clark's discussion of Courbet's "failure" from *The Absolute Bourgeois*:

> A bourgeois artist is shown to fail to make his art "revolutionary," but his failure is in its way exemplary and at least serious . . . it suggests the way in which a struggle against dominant discursive conventions in a culture is bound up with attempts to break or circumvent the social forms in which those conventions are embedded.[40]

Again, a parallel is being made between the artist's work with cultural materials and his recognizing that conventions are bound up and embedded with the institutions and ideologies of the entire social fabric. The formulation here is more carefully posed, but in both passages artists are understood to be engaged in translating resistance between artistic and social forms.

There is another way in which Clark accounts for negation in social terms; he poses the social aspect of negation in a sense that is both more passive and more general, that is, in terms of the social grounding of modern artistic practice. Here, he suggests:

> [Negation] is also identified primarily as a question of the loss and absence of a solid social basis, with practices of negation being predicated on a condition of lack: And surely that dance of negation has to do with the social facts . . . the decline of ruling class elites, the absence of a "social base" for artistic production, the paradox involved in making bourgeois art in the absence of a bourgeoisie. Negation is the sign inside art of this wider decomposition: it is an attempt to *capture* the lack of consistent and repeatable meanings in the culture—to capture the lack and make it over into form.[41]

In a sense, these two social inflections of negation—the one a more passive condition, the other involving agency—seem to echo the modes of casting off in Clark's account of the medium, and can be seen as a social explanation for modernist artistic developments. The point for Clark, which occurs a number of times in the passages above, is always how these social factors translate into, or become manifest as, form. He is concerned with how the techniques and conventions of artistic practice insinuate the specificities and contradictions of their historical locale, and the way that artists respond—consciously and unconsciously—to the uncertainties placed on practice in the modern era. The whole modern problem of the public—imagining what one's art was for, what work it might do, or to whom it might be addressed—is the very ground out of which painterly figures emerge.

Responding to Fried, Clark extends this point: "A strategy of negation and refusal is not an unreasonable response to bourgeois civilization since 1871."[42] Invoking the Commune as a point of historical transition, Clark here draws together notions of revolutionary defeat, the capacity of bourgeois culture to absorb everything into its own terms, and to do so having divested itself of its more manifest bourgeois credentials. This paradoxical social base—or ambiguous ground—opens up a void of signification; art can no longer simply predicate its values on opposition to the bourgeoisie. (There is an interesting twist on this classic narrative of capitalist culture: by the time of writing *Farewell to an Idea*, Clark confesses that he can no longer figure the proletariat as a counterpole of identification, the consequences of which, were we to follow the logic, further compound the problems he establishes here for modern art.) Clark's argument is quite different, of course, from that more familiar life and times of dissident practices of negation where the concept is synonymous with radical, socially engaged cultural activities; this is the version of negation that Charles Harrison dismisses as "the stuff of artistic and intellectual fashion" (although one senses that Harrison includes Clark in his sights).[43] But, then, Clark lets us have this too with "the negation of empty negation," as artists attempt reengagement with their world, and try to challenge the nihilistic implications of modernity and its atrophy of meaning:

> There is a way—and this again is something which happens *within* modernism or at its limits—in which that empty negation

> is in turn negated. . . . For there is an art—a modernist art—which has challenged the notion that art stands only to suffer from the fact that now all meanings are disputable . . . Brecht's is only the most doctrinaire example. . . . Art wants to address someone, it wants something precise and extended to do; it wants *resistance*, it needs criteria; it will take risks in order to find them, including the risk of its own dissolution.[44]

Although there are various pulls on the meaning of "negation," there is no inherent contradiction in these different descriptions of negation's gravitational center; these different emphases are enough, however, to allow Clark's readers to project some distinct desires.

With all the valences at work in his account of art—this shuttling between negation as a property of the medium and as a social process homologized by the formal characteristics of the work; between the withering of art's social base and the troubled attempts to recapture a social role—it should not be surprising that the discussion of practices of negation generated multiple interpretations. Fried assumed Clark's argument to be nihilistic; some took it to be an advocacy for political art; others felt that practices of negation were not political enough. Soon after this discussion, the phrase itself was dropped from his work, its place taken by "resistance and refusal," an idiom used in the essay on Pollock.[45] Although the change was barely distinguishable—resistance and refusal still clearly belong to the family of negative practices—Clark's deployment of the term "negation" transmuted into a new conceptualization. Released of the burdens of describing a category of practice, with all its associated misunderstandings, negation took on a new life, becoming a performative function within the processes of Clark's analysis. It is the consequences of this development that I now want to explore.

On Searching for a Method

The problem of mediation obviously extends beyond the confines of art history and has a long and contentious history; its stakes have always been high. To wonder about the relation of, say, consciousness and its object, a subject and her world, an artwork and its context—or, the different moments and relations within a totality—is to ask some difficult questions. Such questions are methodological, of course, but, as any familiarity with the critical debates of twentieth-century Marxism will show—and especially with those of Western Marxism and the New Left—they are also deeply political, concerning not just the hows of historical interpretation but also those of present contestations and, therefore, the very question of future social transitions; addressing the last, needless to say, proves especially intractable. Dissatisfaction with the lack of mediation, its elisions, lack of success, or impoverished form is a major impetus behind dangerously unwieldy studies like Marx's unfinished critique of classical political economy or Sartre's seemingly unstoppable consideration of Flaubert. Others resort to the pithy, the aphoristic, as though the verbal compression might force the links and relations to show themselves and cause the requisite fusion. Often seen as rejections of the grand Hegelian syntheses, these methodologies of the fragment, I think, try to come at the problem from the opposite direction. *One-Way Street, Minima Moralia, Heritage of our Time*: all attempt to generate a "third meaning" (and then more) beyond the sum of their atomizing parts. It is as if the authors hope that some sort of implied (and open) totality might emerge through the very process of philosophical montage and adjacency. And as with Walter Benjamin's unrealized Arcades Project, the fragments themselves accumulate in files to rival the magnitude of a tome by Sartre or Marx (think of the assemblage of quotations that make up volume 3 of *Capital*). For the most part, recent thought—both in critical theory and art history—recoils from the problematic of mediation. As a thinker who continues to puzzle the issue, and to do so in the context of art-historical work, Clark is unusual and important.

Time and again since the Left-Hegelians in the mid-nineteenth century, mediation has been equated with political conciliation, capitulation, or compromise. The modern rejection of mediation—as one of the terms of dialectics—has been massively overdetermined by reactions to the communist parties in the postwar decades. The Left's turn against "Hegelian habits of mind" (by Althusser, Colletti, the Italian Autonomists, Guattari and Deleuze, to name but a few) has gone alongside, or at least dovetailed into, the wider intellectual revival of Nietzschean thought. The collapse in the authority of the official communist parties for radical thinkers—and especially of any conviction in their capacity to represent, let alone fulfill, any emancipatory project—has had a powerful effect on the direction of critical thought. In art history, as in the humanities in general, the methodological turn commenced with a rejection of ideas associated with orthodox conceptions of Marxism and the associated grand narratives. This turn was accompanied by an increased preference for microstudies, or else in work whose theoretical orientation was framed by aggressive polemics against anything that might be tainted with "totalizing discourse." Perfected through innumerable repetitions in articles and books, and canonized in countless accounts on disciplinary method, this political reaction—whose impetus was sometimes local and autobiographical—generated its own no less totalizing account of the late twentieth century and was particularly important in producing the closure on the issue of mediation. Once every entity was treated as "text" or "discourse," then the relation of subject to object, or text to context—a central concern for Clark's account of the social history of art—could be rendered obsolete as an intellectual problem. Though there is no doubt that there have been many gains from these developments, this has come at a cost. What interests me is less the polemical thrust, than the points of tension with the methods that were ostensibly rejected. Despite the anti-Hegelian mood rife in cultural theory for much of the late twentieth century, one of the things that has dogged critical thinking in recent years is a sort of teleology—repeated in many histories of art history, or histories of cultural theory—where methodologies or "theories" succeed one another, chapter-by-chapter, in an etiolating metanarrative. Whether the favored narrative structure of these accounts is the process of theoretical annexing or conceptual fine-tuning, such accounts present an overwhelmingly affirmative and progressive sense of their own theoretical becoming—all in the service of disciplinary self-confirmation.

Claims to have overcome—surpassed, solved, or bypassed—the problems presented by mediation are, I think, suspiciously glib. We are not beyond the subject-object binary, least of all when a good number of binarisms seem ever more entrenched. Of course, we are running fast and loose with the terms "subject and object," letting all possible concretizations of the terms coalesce into a singular (undesirable) abstract. This is a method whose metaphorical or political elasticity can stretch and slacken to include whatever one wants, but which is so vague that it can never really be addressed. The power is all in the suggestion. Adorno is guilty often enough of employing such tactics, but he also recognizes the horror that the absence of such distinctions might entail. It is more productive to understand the problem of subject-object relations as a dynamic rather than an absolute one, to see those relations in terms of their various transpositions. To grasp relations dialectically is to be already having to think through a performative contradiction: an opposition may be posited, but on the understanding that its grounding is unstable or contingent; the description is acknowledged as inadequate in the very moment that its deployment is an unavoidable necessity.

Clark's refusal to conform to an economy of straightforward binaries has made his writing especially difficult for intelligent newcomers, which is why both Right and Left mistook the project of *Image of the People* and *The Absolute Bourgeois* and why the discussion of negation led to so much misunderstanding.[46] There is a widespread view that sees Clark's primary contribution to art history as introducing social context and concerns with the conditions of production into considerations of art. That Clark famously disputes Greenberg, for instance, does not mean that he pits contextualism against formalism, a view that does injustice to both parties. Indeed, the simplified oppositions of content versus form, inside versus outside, text versus context seem to be precisely what Clark focuses on as a problem. The point is made quite explicitly in the opening chapter to *Image of the People*, in a passage that is often cited as the essence of Clark's social-historical approach:

> If the social history of art has a specific field of study, it is exactly this—the processes of conversion and relation, which so much art history takes for granted. I want to discover what concrete transactions are hidden behind the mechanical image of "reflection," to know *how* "background" becomes "foreground"; instead of analogy between form and content, to discover the network of real, complex relations between the two.[47]

It is the elucidations of these links, the actual connections which bind together art and politics, with which I am concerned, and specifically the problem of their mediatory articulation: the hows, the complex web of relations, and the processes of conversion, the concrete transactions.[48]

Clark's concern to avoid "analogy" and "the mechanical image of 'reflection'"—or, as he refers to it elsewhere, "confident equations," "vague association," or "conjunctural" analysis—provides a persistent preoccupation in his work and can be seen as implying a twofold critique.[49] On the one hand, it involves a criticism of the academic mainstream of art history and, on the other, an intervention into the intellectual Left; indeed, for all its art-historical knowingness, Clark's project has rarely been addressed solely to art historians.[50] For him the problem with making analogies (or equations, or associations, or conjunctures) is that they register neither particularity nor social totality. The inability to situate a phenomenon within the social totality—or the evasion of such questions—seems to be a charge most directly addressed to the discipline in its more traditional forms, and, as we have seen, the explicit disillusion with totality became intellectually pervasive among the new guard. Meanwhile, Clark's critique of the loss of particularity is primarily pitched to the political Left and to social historians. (This doubling of address is a familiar dialectical strategy in Clark's work: the 1981 preface to *Image of the People* is framed by an engagement with the Right and the Left of the discipline; the early version of the Pollock essay is in dialogue with Fried and Serge Guilbaut, with representatives of both formalist and social-historical accounts of art. Attentive readers also find themselves caught within art history's dual legacy, pulled between the traditions of the philosophy of history and those of philological investigation.[51])

Clark was by no means the first to make this sort of intervention. Writing just over a decade before *Image of the People* and *The Absolute Bourgeois*, Hauser likewise argues against the "naïve form of materialism," or "the very crudest type of social history." Socialist art theory, he complains, often replaces "the untenable doctrine of the absoluteness of spirit by the equally untenable absoluteness of matter, one metaphysic by another."[52] Meanwhile in seeking to elaborate the "transmission belt"—here Hauser plays with the term used by Dvořák and Schnaase—between art and cultural history, between art and philosophy, and between art and social conditions, the discipline of art history traditionally resorts to loosely conceived connections and parallels, "unconvincing analogies and confusing equivocations." Too often, social and cultural phenomena, from the economy and social relations to theological and philosophical ideas, are reduced to "a mere backdrop."[53] Antal too criticizes approaches where the connection between social background and art remain "comfortably vague."[54] The echoes here between Clark and the earlier social historians of art are striking. In this regard, at least, Clark's interlocution seems anything but ironic.

On the whole, the prevailing intellectual disposition since the sixties has tended to favor particularity: for some within the European New Left, this offered a way of marking one's distance from Stalinist orthodoxy; for others it meant rejecting Marxism altogether.[55] Marked by increasing anxieties over the implications of the phrase "in the last instance," these disputes concerned questions such as the understanding of the "base-superstructure model," "vulgar materialism," or acts of "reductivism." Engagements with the thinkers of the Frankfurt School posed, for some, the dilemma of identity thinking, interpreted as much with contemporary Leftist politics in mind as with the reifications of capitalism. The volatile cocktail of issues already indicated here still lingers in attenuated forms in even some of the most postpolitical thought. Although his own philosophy was soon to be one of the most prominent casualties of this intellectual shift, Jean-Paul Sartre's voice from 1960 already sounds most of the keynotes of the new mood. Focusing on the version of method on offer from the PCF, Sartre charges contemporary Marxists with being "mechanical materialists" and "poor dialecticians."[56] Their main representatives, he argues, historicize by means of a "system of correspondences between abstract universals," forcing concepts into "prefabricated molds" and a priori models.[57] Thus Sartre insists:

> The open concepts of Marxism have closed in. They are no longer *keys*, interpretive schemata; they are posited for themselves as an already totalized knowledge. . . . The totalizing investigation has given way to a Scholasticism of the totality. The heuristic principle—"to search for the whole in the parts"—has become the terrorist practice of "liquidating the particularity."[58]

The danger is that one might make one's mediations, or interpretive reductions, too hastily and inflexibly; the aim is, rather, "to discover a supple, patient dialectic."[59] Clark's project needs to be seen as being lodged within this problem.

In art history, as elsewhere, the large, sweeping epochal histories—whether called *The Social History of Art* or *The Story of Art*, be they framed by an earlier Marxist or by Popperian preferences—were the subject of much criticism, and attention turned to closely focused studies. Clark's work itself forms part of this general intellectual tendency: a few years in the career of Courbet; the way the transformation of Paris in the middle of the nineteenth century registers in the work of a handful of artists. Even in *Farewell to an Idea* we encounter a small group of exemplary figures—David, Pissarro, Picasso, El Lissitzsky, Pollock, Hofmann—and follow their engagements with the contingencies of modernity, though the synthetic claim of the title belies a deeply fractured story. Particularity was a prominent concern of other art historians of the New Left. In a classic essay from 1972, Kurt Forster also argues against the use of analogy to establish parallels between "art" and "history." He notes how analogies are often wielded in the histories then available as a means to explain those moments of rupture in the standard art-historical narratives. Forster distinguishes a traditional from a social history of art. The former, he suggests, relies on abstract universals, philosophy, and a history of ideas, and posits art as a matter of direct experience and spontaneous access. While to traditionalists, social historians are incapable of experiencing artworks, Forster argues that the social history of art concerns itself with material production and ideology critique, with the concrete and the particular.[60] In another influential article criticizing conventional modes of art history from a Marxist perspective, O. K. Werckmeister similarly counterposes philosophical abstractions to concrete histories, and criticizes the "ideological generalization" of academic art history. Moreover, Werckmeister argues that the same problems of idealism and determinism are replicated in Marx's own approach to art. Compounded by a mood of political resignation, this contradiction, he continues, manifests as an overemphasis on aesthetics as a locus of revolutionary potential, and, methodologically, this situation leads to an unwarranted emphasis on abstract philosophy at the expense of studies focused on concrete specificities and ideology critique.[61]

The question of ideology provided one of the fundamental debates for the elaboration of the social history of art in the seventies and eighties, turning on its peculiar type of mediating function. Werckmeister argues that ideology "projects harmony between two social positions that were actually antagonistic, and in whose continuing conflict it was meant to serve as a diversion."[62] Ideology, in other words, is a false mediator, proffering consensus when really there was dissension. Although not contradicting Werckmeister, Clark gives the issue a somewhat different emphasis, insisting that categories such as "ideology" or "contradiction" need to be brought into conjunction with the concrete particulars of material practices. Here, it is the politics of the particular that is important, not just the more general concerns.[63] In an essay from 1969, he notes how myths themselves are "in a state of constant development and self-contradiction."[64] This point is developed further in *Image of the People*: the bourgeoisie, he argues, eliminated the "middle terms of [the] process, the unstable categories . . . in which bourgeois identity is gradually and painfully assembled."[65] His interest is clearly more concerned with the movements within ideological processes, and how they work into the mindsets of individuals and classes. Ideology, as Clark later reiterates, may be a process that generalizes repressions and imagines "contradictions solved," but it "takes as its material the real substance, the constraints and contradictions, of a given historical situation."[66] The spirit again seems close to Sartre's well-known words: Valéry may have been a petit-bourgeois individual, but not every petit-bourgeois was Valéry. Couched in terms of the problem of the general and particular, or the abstract and the concrete, we will need to tread carefully through the varying emphases invoked by the use of these oppositions.

As Clark puts it, writing in the late sixties, "the facts of discontinuity and contradiction" constitute the object of art history—an imperative that was also directed at the object itself, at the social fabric or at the discursive regime; indeed, for him, discontinuity takes on almost ontological proportions.[67] Courbet exploits the disjunction between bourgeois and feudal popular representations, "images with a dual public, and a double meaning."[68] As itself a form of ideological practice, art history tends to erase the "discontinuity in the object" and to produce a "fiction of a *consistent* and *Real* entity, recoverable in discourse," Clark writes.[69] To counter ideology the historian needs to recover the fissures, gaps, contradictions, and inconsistencies to emphasize the "figures of discontinuity" which damage art history's "assumption of coherence and linearity."[70] We can recognize here the same challenge to Greenberg and Fried; practices of negation are not just procedures used by modernist artists, they also imply a politics of historical methodology. Discontinuity is everywhere: in the object of study, in history, in facts and figures, and in the actions of radical historians. And one hears the ghost of Sartre again: his "supple, patient dialectic" with which he hopes to describe "movements as they really are"; and his refusal "to consider a priori that all lived conflicts pose contradictions or even contraries."[71]

Clark's negotiation of this mood for particularity is, then, distinctive: he seems less inclined to lapse into the routine dichotomies that are still a feature of the practice of criticism. Particularization is not set rigidly apart from the importance of addressing the wider totality; indeed, particularization only makes sense as such through its material dialogue and resistance to its context. Moreover, in contrast to Werckmeister's formulations from the nineteen-seventies, philosophy, aesthetics, or the history of ideas *per se* seem not to be targeted as the realm of idealism nor dismissed as abstract generalizing; philosophy is not simply counterposed to history or politics. Such oppositions have long been a polemical battlefield on the Left, and echoes of Marx's critiques of German Ideology and of Hegel reverberate through academic discourse in its radical modes. There is evidence enough that categories such as "history" and "politics" are no less susceptible to treatment as ideals or abstract universals than are "philosophy" or "theory." Marx makes the point with the more difficult category of "population"—so obviously empirical it would seem—in his *Grundrisse.* There is no consideration of abstraction's concrete immediacy in *Capital* without combining abstract and concrete thinking. We do not need to get lost in the play of abstract concretes and concrete abstracts, or of immediacies and mediated immediacies, to recognize the problems with the usual binaries. To return to an example raised above, although it might be tempting to see the philosophy of history and philology as, respectively, universalizing and particularizing, to do so trivializes a more intricate history. Philological attention to empirical detail does not as such absolve that focus from universalizing actions, as many participants in the Marxist Caucus would have recognized. The issue, then, is to consider the different modes of relations between categories of investigation, between categories posited as the general and particular in the act of knowledge formation.

We find a similar echo in arguments concerning method and the object of study and in the way Clark works the categories of abstract and concrete in his discussions. Writing of Millet's *Ruth and Boaz,* he states: "Almost always it involves a move towards the particular, away from the approximate sublime. Often it means putting things in opposition that started off in harmony . . . . It involves keeping anonymous gestures in tension with actual faces."[72] In a contrasting movement, Clark outlines what he considers to be the problems in Daumier's progression: "The old abstractions of Daumier's art—France, Republic, Charter, Constitution—had worked because they were confronted by the details of a real politics, one he knew at first hand." But without this engagement "the abstractions multiplied and the politics became increasingly second-hand; a politics of 'figures' and 'personalities,' not a politics of the streets."[73] Such tensions, in fact, emerge as a basic problematic for modern art. By the mid-nineteenth century, he argues, the ambiguous relation that artists had with their public was caught between two contrasting dangers: the public had become "either too fixed and concrete a presence or too abstract and unreal a concept."[74] Interestingly, the word "abstraction" in the title "Jackson Pollock's Abstraction"—the original version of *Farewell*'s chapter on Pollock—intimates not just the debates about the artist's drip technique, or those about nonfigurative art. The deceptively literal and banal title also tugs in the direction of abstraction in its social sense. As a discipline, art history is ill equipped to recognize this play of terms. Clark is certainly reflecting on how to think through Pollock's paintings in relation to the reifications of twentieth-century capitalism; Pollock responds to a situation where history is "experienced as a general (abstract) condition or fate," and, according to Clark, "nothing very illuminating will be said unless we manage to enter into the abstract mode and lay hold of its particular ways of meaning."[75] The word "abstract" is doing a lot of work here. It is among the key themes of *Farewell.* But the yarn was been respun; the Red critique of spectacle now entwined with the grays of Weberian disenchantment

As some of the examples cited already suggest, coquetting with the language of Hegel has been one of Clark's particularities, and such play with Hegelian modes was increasingly unfashionable in the seventies, when Althusser's critique of the "immature" Marx gained dominance in the Humanities. In *Image of the People* Clark wryly acknowledges Hegel as one among some "disagreeable topics."[76] Soon after, we find him objecting to the "caricature of Hegel" in circulation.[77] Far from casting Hegel as some General Secretary of Totalitarian Totalization, Clark draws the "dead dog" into a less familiar cause.

In a move blatantly perverse to the usual image, Clark declares his interest in Hegel's "love and labour of the particular." He argues that, with the loss of dialectical thought, the categories of art history became stultified and "incapable of renovation."[78] It is, in 1974, the older intellectual models that still shape the discipline, and Gombrich's perspective also seems to be in Clark's sights. What Clark wants to resuscitate is the Hegelian "habit of mind" or "kind of thinking," with, as he suggests, "its power to open up a field of inquiry, to enable certain questions to be asked."[79] This kind of thought involves "active interplays," the charting of relations that make for living distinctions rather than atrophied oppositions, for dialectic rather than antinomy. He is "for a history of mediations."[80]

A concern over thinking's loss of suppleness and with categorical ossification is often expressed in Clark's writings. In fact, this loss of dialectical fluidity in art-historical reflection is addressed through the example of Erwin Panofsky in Clark's article for the *Times Literary Supplement* and, later, through the example of Greenberg. In the both cases, Clark sees the problems and tensions that once enlivened their work as disappearing from their thought: "Avant-Garde and Kitsch" shrinks into "Modernist Painting" just as the intellectual mobility of Panofsky's "Die Perspektive als 'symbolische Form'" fades from his later writings.[81] The process whereby Greenberg comes to think of the values of art as increasingly self-legitimating amounts, for Clark, to a "ceasing of the dialectic"; Greenberg stops trying to understand the relation of art to society and dislocates its values into a self-contained world. Clark carries this discussion over into his work on Pollock. This time he openly employs Hegelian phraseology, aiming—in contrast to Greenberg—to show the processes of social ingestion, becoming, interiorization and internalization in Pollock's paintings:

> nobody wants the social history of art to be, in Hegel's words, "an external activity—the wiping-off of some drops of rain or specks of dust from the [artistic] fruit, so to speak—one which erects an intricate scaffolding of the dead elements of their outward existence—the language, the historical circumstances, and so on"; but to do more than this is difficult, especially with fruit of this kind.[82]

Just how *do* you mediate between Pollock's art and the social world, Clark wonders, apart from asserting the brute fact of an artist's social and historical belonging?[83] The problem seems particularly acute with a practice like Pollock's. As I suggested at the outset, there is no hiding the problem—that is, no concealing the shortcomings of existing social-historical mediation—behind a version of content analysis or modernized iconography, crutches still available when the object of one's attention depicts a recognizable world of figures and settings to which meanings can be attached. It is not so easy to connect surface facture to an analysis of publics, academies, or everyday life, or, as Clark prefers it, "a certain representational practice inhering in the culture of a class."[84] Whatever the problems of the various accounts of artistic abstraction—whether of the formal or social-historical variety—their methodological difficulties do not threaten to bring those projects to a halt. But when centering mediation as one's main concern, a practice like Pollock's offers scant refuge; it is no longer possible to elide the difficulty of articulating the social-historical significance of an arching line of paint.

Part II

On the Mediation of Metaphor

So far we have been preoccupied with the sort of question that once drew a good deal of attention: how are relations to be articulated across apparent moments of nonidentity without resorting to bald analogy, crude reduction, vulgar materialism, or economism? There is, however, another aspect to the problem, one that is less often addressed: how are relations to be expressed without collapsing all distance and, with it, the ability to make conceptual distinctions? This second pressure, which is just as important and equally as tricky as the former one, is, as Clark puts it, how to sustain a painting's "ordinary, entirely representative distance from the world it is part of"?[85] If this question seems simply to register some degree of art's autonomy, relative or otherwise, it also points to a larger conceptual problem: the terror of identity lurking, for all the talk of difference, in the postructuralist formulation of textuality; it also haunts the formulation—one

Clark adopts from Bakhtin—"context *is* text."[86] This question of distance figures the other half of the mediation problem: without the "gap" between text and context there is only sameness. Clark's point of reference here is *Dialogic Imagination*. However, it may be worth noting Bakhtin's short essay "Art and Answerability" from 1919. Bakhtin emphasizes the difference between internal and external conceptions of unity, which has some echoes of the type of distinctions between symbol and allegory that I will address later, not to mention of Hegel's critique of superficial thought. The passages from Bakhtin also have a lot to say about what would later be called the problem, or dialectic, of the avant-garde—the questions of autonomy and heteronomy, the relation of art to life, and the situation where "life has no hope of ever catching up with art" when the latter is too "audaciously self-confident." Staging little dialogues between "art" and "life," Bakhtin explains how he conceives "inner connection," arguing that it is posed by "answerability," which is not merely a liberal "mutual answerability" but also a "mutual liability to blame" and guilt. Internal unity, then, does not have a plenitudinous ease; its unity "is not one," but a process of uniting that is torn.[87]

The problem of identity and nonidentity can be followed through Clark's articulation of mediation. As one attends to Clark's writings on modernism, what is noticeable is that he does not dispute the notion of autonomy as such, or, to be more precise, he does not dispute it in ways familiar from much of the criticism of Greenberg. Instead of simply countering Greenberg's pro-autonomy account with an antiautonomy thesis, he challenges, as he puts it, "the grounds on which . . . autonomy is secured."[88] In other words, Clark is concerned with the way the concept of autonomy is understood. This tactic is not an unusual one for him to adopt: it seems to be what is at issue in his recasting of the problem at hand, or in disagreements that are "small but definite."[89] Charles Harrison suggests that Clark's account of French nineteenth-century painting "gives flatness a kind of sociological meaning," thereby repositioning Greenberg's account.[90] Harrison is responding to a point in Clark's discussion of Greenberg:

> flatness in its heyday *was* these various meanings and valuations; they were its substance, so to speak; they were what it was seen *as*. Their particularity was what made it vivid—made it a matter to be painted over again. Flatness was therefore in play—as an irre-

> ducible, technical "fact" of painting—with all these totalizations, all of these attempts to make it a metaphor.[91]

The point here is that Clark shifts flatness from a concern purely with medium to a comparison with popular posters and, further, to a homology for the alienation and lack of social depth experienced in modernity. However, while Harrison's formulation is not entirely wrong—and a lot depends on how the word "sociological" is intonated—it fails to do justice to the dialectical movements that Clark sets in play.

Clark's thinking can best be understood through the dynamic conceptions of *Ideologiekritik*, which draw on the logic of commodity fetishism set out by Marx in *Capital*. In bourgeois thought, it is argued, an accurate (mimetic) transcription of reality nevertheless fails adequately to explain or understand social reality. As Marx famously puts it:

> To the [producers], therefore, the relations connecting the labor of one individual with that of the rest appear, not as direct social relations between individuals at work, *but as what they really are*, material relations between persons and social relations between things.[92]

Appearing "as what they really are," these relations nevertheless deflect a full recognition of their own social substance, and are premised upon the abstraction of form from content. In the development of Marx's argument deployed in *Ideologiekritik*, this perception is often reduced to a simple distinction between truth and falsehood: commodity fetishism is read as if it were no more than a mistaken ideology or mystified perception. However, a more attentive reading of Marx would distinguish between the moments of transcription and transfiguration in the way that social reality is grasped by consciousness.[93] Each of these moments may partake of truth and falsity. As representations, both transcription and transfiguration mark their distance from that which they seek to represent; but transfiguration suggests a more substantial action upon the material. In fact, this perspective might make less dichotomous the analytical distinctions made so far in the discussion of Clark's conception of negation, which distinguished between a negative critique that drew its force from the medium and one which did so from factors external to the medium. The Greenbergian claim—that the values of art are self-sufficient, and self-legitimating—are, Clark seems to imply, precisely how things really are, that, if you like, Greenberg and his followers *correctly* transcribe developments in art. But these critics recognize only appearance and miss, or mistake, the substance of these developments—a substance that lay outside the confines of the medium itself: that is to say they misrecognize "the grounds on which . . . autonomy is secured." This structure of argument is also observable in Clark's account of negation, where he writes: "Negation is inscribed in the very practice of modernism, as the form in which *art appears to itself* as a value."[94] The same basic argument underpins *The Painting of Modern Life*: for all his love of the works of art studied, Clark concludes that the painters of the Parisian avant-garde in the late nineteenth century misrecognized the restless characteristics of petit-bourgeois life for the character of modernity per se. Seen like this, it cannot straightforwardly be argued that sociological meaning is ascribed to flatness.

Metaphor is the pivotal concept in Clark's recasting of the problem of autonomy. He uses the term "metaphor" to argue for the sociality of the values established in modernist theory. As we have seen, in "Clement Greenberg's Theory of Art" the category of flatness is seen as emerging from a metaphorical articulation of a number of everyday values current in Parisian life in the late nineteenth century. It is by emphasizing the processes of metaphor that Clark redefines modernist categories; indeed, he suggests, it is by way of the *mediation of metaphor* that the values of art can take the form of, and appear as, being self-sufficient. Clark's emphasis on metaphor directly challenges the attempt to erase tropes from the interpretation of high modernist art. Fried, for instance, following Greenberg, rails against the rhetoric of popular existentialist metaphysics—characterized by loaded metaphors—once current in the Pollock criticism.[95] Indeed, mining such metaphors has become synonymous with the social history of art.

In his fascinating study of abstract expressionism, Michael Leja draws attention to this metaphorically laden corpus of writing, with its webs, labyrinths, and abysses. Leja employs these tropes to reopen the interpretation of Pollock, and this allows him to situate these paintings in the context of what he calls "modern man discourse," a discourse prominent in the United States in the postwar period.[96] Leja—who studied with Clark—notes that his own emphasis on metaphor is a more banal and particularizing interpretation of Pollock than Clark's.[97]

This observation is modestly posed but instructive, helping to clarify the way that Clark uses metaphor in his Pollock essay. Leja adapts the approach deployed by Clark in his earlier writings on Courbet and Manet: reading critical writing for its repetitions and lacuna. Leja focuses on the content of the metaphors circulating in the Pollock criticism of the nineteen-forties. In his own work on Pollock, however, Clark does something very different and his version of the mediation of metaphor operates in another register. In both the essays on Greenberg and on Pollock, Clark notes that for all their hostility to metaphor, narrative, and literary content, modernist critics themselves, nevertheless, tend to secure their interpretations under a dominant metaphoric frame. As we have seen, this metaphor (operating at a meta-level) usually figures the artwork under discussion in terms of plenitude and harmony. Clark introduces dissonance into his account of Pollock not as a straightforward alternative, or counter, to the account of the major modernist critics, but as a way of reframing and complicating their argument. Thus, for Clark, Pollock's practice is seen to oscillate between the metaphoric poles of "totality" and "dissonance."[98] We need to be careful here to avoid reducing the difference between Clark and Leja to a simple matter of high and low content, to a reading where Leja's attention to the popular media's modern-man discourse stands in contrast to Clark's more philosophical terms. Rather, the difference concerns whether to put the emphasis on the content of the metaphors, as does Leja, or to shift the focus, as Clark does, onto their very operations of the metaphoric vehicle. Clark's understanding is concerned with the relay between the meanings suggested by metaphor and the metaphorical vehicle itself. (Unquestionably, "dissonance" and "totality" have metaphorical content, but they also start to describe how metaphor works.) A similar shift of emphasis can also be seen within Clark's own writing. Thus, where metaphor tends to refer the reader to specific meanings in modern life (in "Clement Greenberg's Theory of Art"), its weight (in his essay on Pollock) increasingly pulls away from these particular valences and pushes us toward a consideration of the act of metaphorization as such.

For critics such as Greenberg, Clark argues, the explicit antipathy toward metaphor allowed them to assert medium against meaning, likeness, or literary values. We might add that the high modernist critics considered the eradication of metaphor in the work they valued to have been attained and to be untroubled.[99] Clark sees this hostility to metaphor, however, as revealing the very opposite; on the basis of a symptomatic reading, their antipathy is taken to emphasize the priority of the metaphorical relation itself: "In order to represent at all, I suppose, a series of marks in a picture have to be seen as standing for something besides themselves," Clark writes, "they have to be construed metaphorically."[100] The self-sufficient status of the artwork in high modernist accounts is revealed as a presenting symptom, as nothing but the product of its very social belonging. For Clark, these critics' explicit antagonism toward metaphor—or, at a different level, to nonaffirmative metaphors—was not just a misrepresentation or imposition; he sees the resistance to metaphors, or the "work against metaphor," as substantive to modernist practice.[101] For him, the defining component of modernist practice at this time was its attempts to *negate* its metaphorical status: to find "what it is that stands in the way of likeness." Pollock's marks seek "ways of circling *around* likeness, ways of looking for likeness on the other side of resemblance. And not finding it."[102] There are, however, ambiguities in Clark's presentation of the critics' positions, and he tends to read abstraction as "work against likeness." It is clear, for example, that Fried clearly distances his analysis from a simplistic opposition of representational and nonrepresentational art and prefers the distinction of figurative and nonfigurative, which he considers to be "a more fundamental issue."[103] What seems to be at stake for Fried in the term "figurative" is not the issue of "likeness" but that of figure-ground distinctions. Thus, in his account, Fried can readily identify paintings that are simultaneously abstract *and* figurative; in his writing of the sixties, he lends his critical support to those works that he considers to be both abstract and *non*-figurative, that is, for works that tend to dissolve or complicate the difference of figure and ground.[104] Clark admits to being on speculative terrain. With *No. 1, 1948* in mind, he wonders what is it for a handprint to be seen not as a sign for a figure's hand or for hand-like qualities—or even as an index of some human presence or of the artist—but to be seen as a hand? Figurative, or quasifigurative, painting seemed hackneyed for certain artists at Pollock's historical conjuncture, no longer convincing or able to convey a "relation to the world."[105] To revive this relation to the world, painting needed to reinvent itself fundamentally, and Clark describes Pollock's search for the "origin" of the sign, the "first moment" of the "first" metaphor: "painting had now to find its way back to the ground of representation, to the moment when marks first stood for things other than themselves."[106] Metaphor, then, is foregrounded as the very condition of representation; the processes of representation, it would seem, necessarily emerge *from* the state of metaphor. Without metaphor we cannot imagine representation. The shift from plural to singular—from resistance to particular metaphors to that of metaphor per se—is significant for Clark's sense of Pollock: for a while at least, Pollock's work against metaphor tipped into an effort to negate the metaphorical status of his work. Greenberg and Fried, then, were correct to find in works like those of Pollock a refusal of metaphor, but, according to Clark, they failed to grasp its full (social) significance.

It goes without saying that Pollock's strivings to resist metaphor were exercises in impossibility; metaphoricity cannot be eradicated when it is the condition of one's activity. As Clark notes, "Metaphor is inescapable, and what in any case would an exit from it be like?"[107] Indeed, he argues, in the end Pollock's work against metaphor could only be achieved by pursuing its opposite, by multiplying the metaphors—a multiplication generated, Clark suggests, between those poles of totality and dissonance touched on earlier.

> Pollock's project, as I see it, is exactly *not* to allow the figures of dissonance to hold sway, any more than the figures of totality. His painting is a work *against* metaphor, against any one of his pictures settling down inside a single metaphorical frame of reference. He wishes to cross metaphors, to block connotation by multiplying it. He intends so to accelerate the business of signifying that any one frame of reference will not fit. Figures of dissonance cancel out figures of totality; no metaphor will get hold of this picture's standing for a world, though we think the picture does somehow stand for one: it has the requisite density.[108]

This is a brilliant account, but we should note how Clark avoids making a direct, and all-too tempting, analogy between social contradiction and the metaphors of totality and dissonance. Instead, these metaphoric poles translate into the problems of language and the categories of thought—and Pollock's struggling with the "grounds of representation." The casting of Pollock's practice as a work against metaphor, then, serves to move that practice onto the terrain of aporetic activity: as Pollock is caught in the economy of his own denial and multiplication of metaphor, the insecurity and continual restlessness of this irresolvable contradiction underpins his project.

On the "Permanently Unresolved Dialectic"

In Clark's essay on Pollock, this aporia invokes the dilemmas of the institutional recuperation of any critical move and fears about the futility of practices of resistance and refusal. Clark terms this fear "the bad dream of modernism." These are familiar concerns, but the sense of impasse in Clark's essay also has a further characterization. As we have seen, the substance of Pollock's practice is described through the contradictions of, among other things, metaphorical denial and multiplication, the latter operating through figures of dissonance and totality. I have noted a shift of emphasis in Clark's method, but this should not be seen, as some critics have been inclined to do, as a move into deconstruction or textualism. The role of representation is fundamental to his project, which is one reason that his work simultaneously keeps its distance from both poststructuralist and reductive social accounts. The point here is that Clark's engagement with metaphor casts the very mechanisms of representation onto a socio-political terrain. Aporia is foregrounded as a question of history, as a problem of representation within capitalist culture, precisely because Pollock's paintings are considered as "standing for a world." As Clark puts it:

> Pollock's painting in its best period . . . is contradictory; it lives on its contradictions, thrives on them, comes to nothing because of them. Its contradictions are the ones that any abstract painting will encounter, as long as it is done within bourgeois society, in a culture which cannot grasp—for all its wish to do so—the social reality of the Sign. That is to say, on the one hand, abstract painting must set itself the task of *cancelling* Nature, ending painting's relation to the world of things. It will make a new order to experience, it will put its faith in the sign, in the medium, it will have painting be a kind of writing at last, and therefore write a script none of us have read before. But on the other hand, painting discovers that none of this is achievable with the means it has.[109]

From this perspective, the delectable impasse of *Lavender Mist,* to which Clark refers in his response to Fried, becomes a metaphor—albeit less delectable—for the status of representation or language under bourgeois social relations.[110]

In this argument Clark draws on Hegel's concept of the unhappy consciousness to figure his sense of impasse. Migrating from the footnotes of the essay version to become the chapter title in *Farewell,* this concept is no minor player in—or, indeed, a mere distraction from—his argument.[111] It is not, as he emphasizes in *Farewell,* to be reduced to "a catchphrase summing up Pollock's method, or modernism in general," but Clark remains committed to the substance of his earlier formulation: what Hegel has to say on the unhappy consciousness is, for Clark, no less than "the essential framework for an understanding of modernism and its permanently unresolved dialectic"—and, in particular, for understanding Pollock's practice.[112] The unhappy consciousness—which features in the *Phenomenology of Spirit,* shortly after the section on the master-slave dialectic, indeed, as the introjection of the dynamics of the dialectic of lordship and bondage—is defined as "the consciousness of self as a dual-natured, merely contradictory being."[113] However, the awareness of self-contradiction is limited, and, as Hegel puts it, the unhappy consciousness "is not as yet explicitly aware that this [duality] is its essential nature, or that it is the unity of both [parts]." Hence, the unhappy consciousness is a consciousness built on "the positive moment of practicing what it does not understand."[114] This play of identity and nonidentity generates a continual striving for the reconciliation of one part with its other—a striving to unite with or destroy that otherness—that is, in truth, already itself. In a sense, the unhappy consciousness is aporia par excellence—antinomy, impasse, and difficulty. Perhaps more suggestively, however, it implies a persistence, a continual going at, or over, a problem—and does all this despite half-knowing the impossibility of solution or end (at least, in the terms at its disposal). Seen thus, the dynamics of the unhappy consciousness are more far-reaching, more intractable and historical, and more involved than those of its apparent close relative, undecidablity.

Modernism's permanently unresolved dialectic, then, is Clark's final resolution, or account, of Pollock's art. Indeed, in this essay Pollock's *The Wooden Horse* of 1948 (see figure 1.2) comes to figure bourgeois culture's inability to grasp "the social reality of the Sign." Its awkward attachment of the head of a toy horse onto a ground of loose cursive skeins of paint

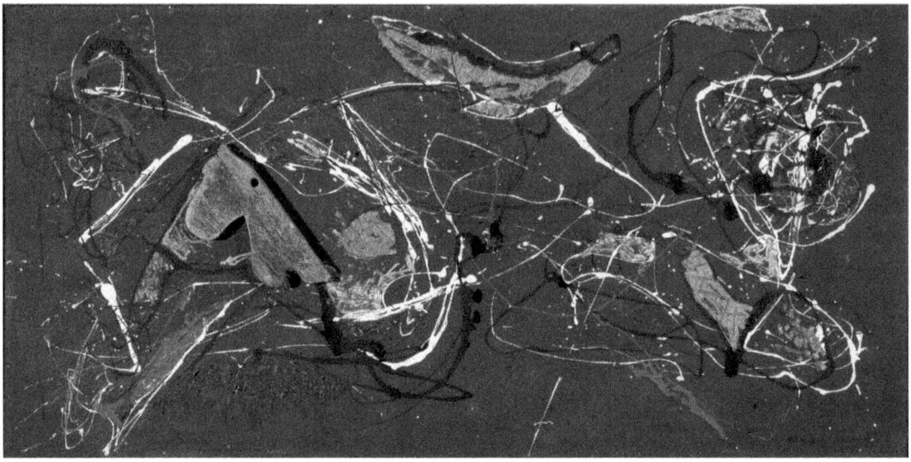

1.2 Jackson Pollock, *The Wooden Horse: Number 10A*, 1948, oil, enamel, and wood on brown cotton canvas mounted on board, 90 × 178 cm, Moderna Museet, Stockholm. Photo: Moderna Museet. © The Pollock-Krasner Foundation ARS, NY, and DACS, London 2009.

supplants the impasse of *Lavender Mist*. Fried charts a similar contradication in Pollock's work. In Fried's account, Pollock tries "to achieve figuration" and "to restore" a traditional figurative role to line but within an overall project that renounces figuration (that is, in favor of the "overall, optical style"). Pollock hopes to do the (seemingly) impossible, Fried thinks, a contradiction that, he insists, we need to grasp "as vividly and even as painfully as we can."[115] Clark takes the same contradiction to be a social question of representation. Faced with the aesthetic unease of the wooden head—its failure to resolve into a picture—Clark imagines Pollock's response to the irresolvability:

> So let the Figure be in the picture on purpose—in the negative, taking the lordly, footling, infantile form that is the best (or worst) that painting can do with it. Let it be there *as* negation, as the sign of antinomy, not dialectic. For the grounds are lacking on which the contraries of bourgeois art—its claim to Nature and its wish for the free play of the signifier—could be dialectically reconciled.[116]

Both sides of the antinomy of bourgeois art—the claim to an identity of sign and referent and the claim to absolute nonidentity—are attendant upon, or articulate, the work against metaphor. For Clark, then, the permanently unresolved dialectic of modernism is the problem of *bourgeois* society and *its* inability to recognize "the social reality of the Sign."

Our difficulty comes, though, in thinking through the consequences of this recognition (and the recognition of its failure); in what sense might the sign and the social, or text and context, be reconciled? Might some postbourgeois society have a capacity to articulate those mediations that would allow, say, a drip of paint to appear "social" and do so as any more than mere brute fact or sheer assertion? But in this reading of Clark, the inability to grasp the social constitution of representation refers not to an incapacity to imagine the future but rather to the inability to see what *already* exists. Moreover, this problem is not unique to Pollock, bourgeois society, or modernist art practice or theory. Never mind bourgeois blindness, even those who do "recognize the social reality of the Sign," who actively seek out the mediations, are caught in the same problem, suspended between their (theoretical) recognition and the inability to pierce its reified and fetishized forms. The motif of the unhappy consciousness, then, pervades Clark's work in two ways: increasingly as the object of his deliberations and also as a continuing condition for them. This condition comes increasingly to be seen as the problem of interpretation itself. And it starts to live out its own delectable impasse. The problem of metaphor as a mediatory vehicle insinuates itself into the heart of Clark's analysis. The discussion of Pollock's aporia, in which the negation of metaphor unravels as the multiplication of metaphor, unfolds internally through Clark's own method, where his search for mediation returns to disjunction. It is to this second condition, then, that I now turn.

On Anticipation

To enable a way of thinking beyond the conceptual antinomies (inside/outside, before/after), and to avoid treating text and context as external oppositions, Clark draws on Bakhtin's social, performative and dialogical account of language: the utterance is always made of contextual material; text is always already bespoken. This observation provides Clark with a model for thinking not only the textual inside and contextual outside of Pollock's painting, but also its temporal constitution. An important concept derived from Bakhtin's dialogism is "anticipation":

the way that an utterance expects a response and is pitched into an arena known to be already "charged with value."[117] In making this comparison, Clark temporalizes text and context, treating their relation as dynamic and processual, and invokes the concept of the artist's "imaginary public."[118] However, as the argument develops, we find that we are left with a rather uneasy formulation; what proved a subtle reconceptualizing of an artwork in Clark's discussion of Courbet—the painting as a condensation of complex processes and social dialogues—becomes, in Pollock's "specially elaborate, pondered" case, a more reduced idea. The "anticipation" is of the bad dream of modernism, and specifically, the nightmare signified by Cecil Beaton's use of Pollock's paintings as backdrops for a fashion shoot. The topic of Beaton emerges within a rhetorical exchange, so Clark is, I think, using Beaton as a cipher for broader problems, essentially the problems presented by a tension between critical practice, on the one hand, and commodification and appropriation, on the other. But in the process of doing so, too much specificity condenses around Beaton as such. While Clark prefers not to see in these photographs the transmutation of an aesthetic object into mere everyday commodity, he nevertheless invokes one of the major anxieties in the discourse of mid-twentieth-century American art. Both Rosenberg's infamous comment about "apocalyptic wallpaper" and Greenberg's efforts to drive decorative tendencies from abstraction—even Adorno's observations on the paintings used in hotel rooms—registered this concern that abstract painting came perilously close to ornamentation.[119]

We are left, nevertheless, with a rather grand claim for abstract painting, one that puts it into a dialogue on a historical scale. Indeed, if Beaton is placed as Pollock's prescient or phantastic public—albeit as a negatively defined and drastically withered one—we find that this public is also "too concrete a presence."[120] Or, are we to think that by the mid-twentieth century even an imaginary public is hypostatized as an abstraction of exaggerated proportions, even as an absence? Beaton clearly encapsulates a problem within Clark's account, yet as a figure he seems too fixed and particular, while the bad dream becomes just too abstract. Moreover, if we continue to think through Clark's larger project, we must also wonder how this bad dream is interiorized; in other words, what are the means through which this anticipation inhabits the painting? In the process, Pollock's art is made to prefigure history. In certain elements of the radical aesthetic tradition—Ernst Bloch, would be a prominent example—prefiguration, through the anticipation of the future possibility of subject-object reconciliation, acts as both a reminder of its current impossibility and a Utopian sign. Clark's version of anticipation, however, prefigures not some radical future of a different social order (mediating such desires in or through art) but a rather more immediate tomorrow—a tomorrow of continuing unhappy consciousness, of continuing alienation of representation and reality. Embedding its social belonging through dialogism's anticipatory processes, Pollock's abstraction prefigures nothing more that the social constitution of actually existing capitalism. Clark's temporal inflation of the dialogic concept—where anticipation moves from an utterance between two speakers to Pollock trying to outmanoeuvre all that Beaton might represent for his historical moment—is matched by the temporal compression of prefiguration. The device that enables Clark to articulate a painting's inescapable sociality also ends up as a form of historical closure.

It is quite possible to read all this—whether Pollock's practice or Clark's interpretive framework and narrative devices—as an allegory for a historical situation, marked by the aspirations and disappointments characteristic of various generations of radical intellectuals: the fallout post-Popular Front and post-1968, one that pans back and encompasses other lost moments, 1923, 1917, and 1871, arguably even 1848. If this observation reeks too much of vulgar reduction, equally, it seems absurd not to recognize the spiraling potentiation of revolutionary defeat and historical impasse into thought. We have here a version of Benjamin's dialectics at a standstill—a figure that, incidentally, has been compared (both approvingly and disapprovingly) to the patterns of the unhappy consciousness. The social hypostatization of Clark's Pollock echoes Benjamin's melancholic or, alternatively, Adorno's Schönberg. Yet Clark's decision to have the wooden head figure "as negation, as the sign of antinomy, not dialectic" suggests yet another sense of impasse or disjunction.

It is useful at this point to return to the questions of metaphor and mediation. The character of the word "metaphor," both in Clark's texts and in this account, has been loose and generic. For Clark it signifies not a particular trope but an encompassing term for tropes in general. Thus, Clark insists, in his dispute with Benjamin Buchloh over the legacy of Duchamp, that the index is not distinct from metaphor but simply one of its forms.[121] Nevertheless, it may be useful to probe the issue a little further. In detailed tropological studies, metaphor, understood in its specific sense, has a double function, enacting a "transfer" of similarity and difference.[122] This dual operation may help to explain how, while it provides Clark with the means to understand how text and context are mediated, the concept of metaphor seems at some deeper level to undermine this very mediating role by effecting an inversion. Where he starts by asking questions of the relation of art to the social (*how* does background *become* foreground?), he ends up with a moment of nonrelation reinstated (let it be there as antinomy!). The aporia in Clark's object (the point where Pollock's negation of metaphor unravels as the multiplication of metaphor) finds an equivalent in his own methodological project (the search for connections unravels as the return to disjunction). This disjunction emerges in two ways: it is posed as both the historical impasse of our time (the failure of social revolution and the loss of agency) *and* as a fundamental moment of nonidentity in representation that cannot be fully mediated, for all the wish to do so. As already suggested, these two senses are all-too-often ritually counterposed as History versus Language. Accordingly, for those who emphasize the latter term, the dynamics of the unhappy consciousness show a refusal to accept the resistant character of language; for the advocates of the former, it is only the condition of language within a specific social and historical situation. Clark touches on this difference, by way of a discussion of Bakhtin and Paul de Man; as he observes, for de Man "the Unhappy Consciousness is entirely correct in its view of mind and its objects, and that any attempt to think the question of otherness and utterance further—further than its antinomies—is no more than a dialectical sleight of hand."[123] I will come back to de Man and the unhappy consciousness in the consideration of allegory. Posed as antinomies, and in caricatured form, the difficulties of thinking through this issue are evaded in a process where, to use Fredric Jameson's formulation, "one reified theme . . . [is opposed] by another."[124] It is an open question whether or not it is the trope of synecdoche or that of metonymy that best seems to sum up Clark's initial attempts to find the connections between text and context. Too often, though, discussion is closed in efforts to view an entire oeuvre through the frame of a single, dominant trope or by approaches that adopt the "good trope/bad trope" method of analysis.[125] It would be difficult to trace social articulations of cultural phenomena without recourse to tropological expansions and transcodings between different fields of discourse; what is primarily at stake in the deployment of different rhetorical figures, then, are "the dynamics of the trope" and the specific kinds of "reductions or integrations" they imply.[126] The point is not just that Clark favors particular tropes or figures, but that there are movements between subject and object that result from pursuing a dialectical interpretation.

The difficulties discernable in Clark's search for a method ought not to be taken as negative judgments nor as criticisms simply local to his work. Some commentators maintain that the unhappy consciousness is a pseudo-dialectic; others see it as thought's necessarily aporetic state; there may be, then, some fairly intractable difficulties at work here. Clark is, it seems, well aware of the problems. He has remarked that we will not ever magically escape the metaphorical divisions of inside and outside, text and context, before and after, and so forth.[127] To put it another way: we cannot escape those very divisions that Clark's project for the social history of art had hoped to avoid. The difficulty of articulating the mediations, then, is really the difficulty of speaking without recourse to—or to be more precise, without the fossilization of—those analytical oppositions; the need to use the terminology may be inescapable, but, though forces push us there, the petrification of categories is not, and it is the latter of which we must beware. The problem I have been scrutinizing in Clark's work is, perhaps, the inevitable outcome of confronting the difficulties of mediation and making them "into the motor of the argument."[128] Writing in 1994, he acknowledges the demise of the social history of art as a research program, while still affirming his commitment to "the *kind* of questions" it posed.[129] In his 1974 article for the *TLS*, for all his apparent optimism about the possibilities of the Hegelian habit of mind, Clark remarks that a turn to dialectics will provide no simple panacea. What such a turn can do, he suggests, is bring us back into critical dialogue with what he calls "the fundamental questions": namely, "the conditions of consciousness, the nature of 'representation' ".[130] We will, he says, be on a "territory beyond ideology" where we will be susceptible to the "old concepts."[131] There are echoes here of Hauser's discussion of the limitations and possibilities of the social history of art, which "does not work miracles or solve all problems," his reservations about "the classical formulations of [dialectical] theory," and his regret that he did not—in his *Social History of Art*—exercise enough caution in applying dialectics to the study of artistic objects.[132] Clark does not provide details of the difficulties that he has in mind, but he seems to point us to the problem of our categories of thought. Perhaps the concern is with the nonidentity of consciousness and being, thought and reality—what Lukács refers to as "the unbridgeable abyss between concept and reality" and the "'pernicious chasm' of the present."[133] Hauser even argues that "the translation of economic conditions into ideologies remains a process that can never be completely clarified; at some point or other, it involves a gap or leap." Such discontinuities of interpretation are unavoidable, the condition of understanding and the result of the necessary abstractions of thought: once nature is "left behind, we do not anywhere encounter the merely material; we may think we are talking about material conditions, but the leap into the realm of spiritual conceptions has already occurred."[134] Discussing the process of Becoming, Lukács notes how "every attempt to overcome the duality dialectically in logic . . . is doomed to failure."[135] Of course, on this last point, Lukács poses the matter in much stronger terms, for he is speaking of "a system of thought stripped of every concrete relation to existence"—not an accusation one would want or even be able to throw in Clark's direction, but we are back at the heart of what it is to interpret a historical artifact or trace, and to the binds on that process.

The work of the Understanding, Hegel writes in paragraph 32 of his *Phenomenology*, is the activity of dissolution, the dissolution of an immediate relation, and the breakdown of elements or moments. The acquisition of some freedom of these elements from one another, he argues, reveals "the tremendous power of the negative": such is its power of "devastation" that "death" is threatened. Immediate unity is destroyed in order for Understanding to progress, but that moment of devastation risks losing the idea to the unreal. To step back from this risk, however, entails a greater threat:

> But the life of Spirit is not the life that shrinks from death and keeps itself untouched by devastation, but rather the life that endures it and maintains itself in it. It wins its truth only when, in utter dismemberment, it finds itself. It is this power, not as something positive, which closes its eyes to the negative, as when we say of something that it is nothing or is false, and then, having done with it, turn away and pass on to something else; on the

> contrary Spirit is this power only by looking the negative in the face, and tarrying with it. This tarrying with the negative is the magical power that converts it into being.[136]

In Adorno's view, the promise of Hegel's "program of self-yielding" to the object is never realized because "Hegelian thought finds satisfaction in itself." As he went on to argue, "immersion in particularity, that extreme enhancement of dialectical immanence, must also be the freedom to step out of the object, a freedom which the identity claim cuts short." In one of his most memorable passages, Adorno describes how "if thought really yielded to the object, if its attention were on the object, not on its category, the very objects would start talking under the lingering eye."[137] It is also Clark's hope that visual tarrying—a process given full reign in his study of Poussin—might bring the object of his attention to life. Out of lifeless immobility—both the sheer staticity of artifacts and the reified forms of social being and consciousness—might emerge gesticulation, speech, even poetry or a melodious chorus.[138] It is worth returning, at this point, to Clark's comments on the distinction between Pollock's "work against likeness" and "work against metaphor." The dialogue, as we have seen, is with Fried; and as we have also seen, there are problems in the use of the word "likeness." Clark argues:

> A painting could be freed of all traces and afterimages of likeness and still not do the . . . work against metaphor. . . . We could have—no doubt we do have—a strictly "optical," nonfigurative abstract painting that nonetheless stood in a *confirming* relation to a world we might recognize—an inert relation—a relation to "Nature," say, in which nothing of that dismal category was in the least negated.[139]

The implication of Clark's statement is significant. The work against likeness is a project that may be seen through to success and still retain an affirmative relation to the world. The "work against metaphor," however, does not necessarily—or, to put it in stronger terms, necessarily does not—stand "in a confirming relation to a world we might recognize"; it opens onto a deep sense of negativity, and its unsettling is a radical one.

It is interesting, then, to note how allegorical figures come to haunt Clark's work. There is no mistaking the interest in the mechanical "grating and locking" in the essay on "Freud's Cezanne," where Clark once again aims to explore "the system's antinomies and blank spots" and "the powers and limits of a particular system of representation."[140] Nor do we miss "vulgarity," the key category for "In Defense of Abstract Expressionism."[141] The essay on Pollock stages psychomachia, like those of old-fashioned Virtues and Vices, but whose vehicles now carry the attributes of modernity. History, Nature, Events, Sign: even with their capitalization removed (as they are in *Farewell*) these terms tend toward traditional abstract personifications while holding back from any clear delineation of character. Unlike Lucien de Rubempré or Theodose la Peyrade, the provincial aspirants, a petit-bourgeois Thuilliers, the high-financier Nucingen, or even the strained demonic caricatures of a Cerizet or Jacques Collin—with their typifications oozing out of their urban fabric—not even Lukács's Balzacian realism can "get hold of" Clark's figures. We are, then, back with Jameson's reified themes—appropriate, maybe, when abstraction itself falls under a different historical sign, with forms that are dramatically potentiated.

Seen from this perspective, Clark's founding problematic for the social history of art—mediation—may be less critically stymied by abstract art than by the exacerbation of social hiatus, one powerful enough to rupture the revolution of modern art from the modern art of revolution. To renegotiate the abyss between thought and objective reality navigated by Hegel's philosophy, Lukács marshals social agency—the proletariat—a strategy that, for today's intellectuals, Clark included, adds yet another circle to impossibility and another ring to consciousness's unhappiness. His extended leave-taking in *Farewell* is at once testament to a negative dialectic of meaning and interpretation and to a catastrophic historical defeat. If Clark's work is marked by aporia, this is not to dismiss the project triumphantly. Quite the contrary. His ruminations—worked through and tested against his materials—occupy a compelling position in a debate more often prone to hackneyed or commodified gambits. Mediation is not reduced to "the will to reconcile"; it serves the opposite cause. Drawing on the mechanical figure of allegory does not demand a denunciation of organicist dialectics; it becomes, instead, the very problem of the latter's tensile relation with, and historical rendition as, the former. One is dealing, of course, with old concepts, but in new forms and for a new moment. Ultimately, it is here, then, where Clark's dilemma takes us—and, rightly, he provides no pat answers or consoling words. But, he insists, "the survival of the Left matters greatly."[142] One is tempted to say, at least for the meantime, let this troupe of shriveled abstractions march on as a reminder of the historical problem before us.

Chapter 2

Looking the Negative in the Face

Manfredo Tafuri and the Venice School of Architecture

> Already the Great Khan was leafing through his atlas, over the maps of the cities that menace in nightmares and maledictions: Enoch, Babylon, Yahooland, Butua, Brave New World.
>
> He Said: "It is all useless, if the last landing place can only be the infernal city, and it is there that, in ever-narrowing circles, the current is drawing us."
>
> And Polo said: "The inferno of the living is not something that will be; if there is one, it is what is already here, the inferno where we live every day, that we form by being together. There are two ways to escape suffering it. The first is easy for many: accept the inferno and become such a part of it that you can no longer see it. The second is risky and demands constant vigilance and apprehension: seek and learn to recognize who and what, in the midst of the inferno, are not inferno, then make them endure, give them space."
>
> —Italo Calvino[1]

Part I

With damage to a lagoon's ecosystem caused by nearby petrochemical plants and the disturbing frequency of *acqua alta*, Venice's environmental problems are familiar to its visitors and taken to be a pressing issue for world heritage. In 1999, the then mayor of Venice commissioned Oliviero Toscani—the figure famous for initiating Benetton's controversial advertising strategy—to produce a poster campaign intended to discourage a certain kind of tourism. It was argued that the mass of day-trippers—those who stay in the city for just half a day, en route to an even briefer stop in Padua—were draining, rather than contributing to, the local economy. The images Toscani produced to put them off visiting Venice showed rubbish, dead pigeons, two-hundred rats in the Grand Canal, and dogs mating on St Mark's Square. Whether this negative publicity campaign could ever hope to undermine the coach-tour industry is not clear, but it certainly made its mark on the history of advertising. Marketing has long been seen as the classic mode of appropriating avant-gardist negation—and as the very proof of the avant-garde's incorporation—and Toscani has been widely taken to be the commodified inheritor of *épater le bourgeois*. Unfortunately, the litany of vermin upset many local investors, those who the mayor had hoped to support. How could anyone have imagined that the message would reach that niche of peripatetic day-trippers while leaving untouched those tourists whose presence was seen to contribute to the city's economy: those who stayed a night or more, who paid admission fees to the *scuole* to admire the works of Carpaccio and Tintoretto, or attended the Biennale or Mostra, sampling the bars and restaurants? In the context of the environmental crisis facing the city, however, some accepted that shock tactics were necessary. And, as if to raise the stakes of controversy—and as though recognizing the contemporary limitations that applied to even his own shock approach—Toscani had allegedly included a poster showing two men kissing. This conflation and association of homosexuality with urban infestation had the power to disturb even that most jaded and experienced ad reader, unmoved—and even mildly amused in a distracted sort of way—by images of canine copulation. As I say, the whole thing could never have been expected to reach its ostensible target, which is to say that the target was probably elsewhere, among those in (or critically aware of) the advertising industry itself. All par for the course at the end of the twentieth century. Yet, while not divorceable in any simple way from Toscani's operation, the strategy of Mayor Massimo Cacciari carried some rather different resonances. Cacciari was, after all, no ordinary mayor. His willingness to let the very force of the commodity rip into commodification—as if unleashing an autoerotic, self-consuming energy—and the intention to interrupt one set of industrial flows by way of another, echoed some of his earlier political commitments. In this risky play with a strategy of Nietzschean nihilism—albeit one appropriated for a radical-Left project—the philosophical discussions and debates of the seminar room were put into practice at the level of local administrative politics.

Completed nihilism proposes arguments that have been controversial enough in theory, but Rem Koolhaas—who made a name for himself as a critical architect—introduced the problem into practice at

Euralille, the hub for Eurostar and TGV railways developed in the northeast of France in the old industrial city of Lille. His team, the Office for Metropolitan Architecture (OMA), who had responsibility for the overall development, directed—and deliberately miscast—a number of big-name architects. Jean Nouvel, an architect usually associated with exquisite, prestigious, high capital-value projects, for example, was asked to design the cheapest structure, the indoor shopping center. But Koolhaas tried to push the point further, with OMA's own contribution: the "Espace Piranesien." This was OMA's code name for their intervention at the heart of the Euralille project; instead of adding a structure, OMA extracted mass to reveal "the highway, railway, three levels of parking, and the metro, which dives underneath the whole complex, in one overtly metropolitan moment" (see figure 2.1).[2] Bringing the traffic into tense proximity, there was no softening, no masking, no postmodern adornment but instead, raw modernity, unadorned concrete, multistory car-parks, elevators and escalators, walkways and dead spaces, all collided into a disorientating and vertiginous space linking railway terminal with metro, with the local motorway racing along one side (scarcely shielded, visible, audible, breathable)—an abyss

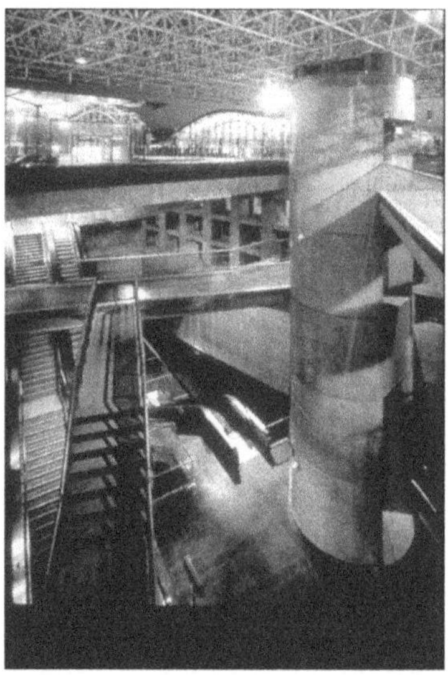

2.1 Office for Metropolitan Architecture, view of the *Espace Piranesien*, completed 1994, Euralille, Lille, France. Courtesy of Office for Metropolitan Architecture.

of dystopic display. In his account of the project, Koolhaas revels in the "infernal complexity" of the whole Euralille project, and seems a little surprised that his team was allowed to proceed with their demonic proposals.

Subsequently, he returns to interview Jean-Paul Baietto—"the father of the plan"—to find out why he never refused OMA's "most outrageous suggestions." Baietto replies that, at this point in history, a number of conditions are necessary in order "to create something worthwhile": clear limits, external demand, and a third term. As Baietto describes:

> With these two conditions, you create the third: you establish on your domain a *dynamique d'enfer*, a dynamic from hell . . . So complex become all the interconnections, the mutual dependencies, the proliferation of interfaces, the superimposition of users and owners that together they form a group of prisoners, shackled by mutual obligation, exacerbated by the very complexity that you offered unwittingly.[3]

Baietto is presented by Koolhaas as the ultimate puppet master, as a Mephistophelean character who, all along, is able to see well beyond and through the architects' own fantasies of control; Baietto reveals, as Koolhaas puts it, "that our seemingly spontaneous action had been nothing but a figment of his imagination." And there, on the page, sits a slightly disheveled and besuited Baietto, pipe in mouth, elbows resting on an invisible table with his hands clasped, looking straight out at the reader with a supercilious and wordly air. The first demand (the territorial limits of local specificity) drains away: we too, wherever we may be, are caught in this man's sights. But of course, in OMA's canonical *S, M, L, XL*, we are part of Koolhaas's rhetorical game; this photographic homunculus is inserted into the spread, his interview staged. Baietto may do little more than reflect back, in terms both elegant and thoroughly disenchanted, the architects' own Piranesian metaphor (or "code"). Koolhaas seems satisfied with this hall of mirrors, reflecting (and inflecting) yet again Baietto's statement:

> It is only when they are all tied to the site by each other's demands, chained together by an overall vision never entirely revealed, when the dynamic from hell makes the entire situation

> irrevocable and the project is like quicksand from which no one can escape, that you can get away with such an enterprise in Europe.[4]

Koolhaas's discursive playfulness—and the nature of Euralille as a space at the heart of new Europe and international travel—makes this void far less troubling than the repressive face of modernity soon to be experienced along the track at Sangatte. Even in the Espace Piranesien, the sound of the cars and trucks flashing by behind a transparent screen is a little less fractious than Godard's comparable cacophony at the start of *Weekend*. Among the options available while waiting for a connection at Euralille, the Piranesian space, which is free of cafes and sales outlets, is a rather interesting place to pass some time. All in all, a rather benign outcome for completed nihilism.

Both Cacciari's project of negative publicity and Koolhaas's Piranasian space can be seen to have their origin in the intellectual and political strategy developed in the sixties at the Instituto Universitario di Architettura di Venezia. IUAV has a sort of mythic presence in modern architectural education. Along with the Bauhaus, the Chicago Institute, Black Mountain, or the Fine Art courses at Yale or St. Martin's, the influence of the Venice school of architecture has resonated well beyond those who passed through its doors. Acquiring its status as a radical institution for the study of architecture under the directorship of the architect Giuseppe Samonà from 1945, IUAV employed, among others, the architects Aldo Rossi, Carlo Scarpa, and Carlo Aymonino. Manfredo Tafuri, a historian and theorist of architecture, was appointed in 1968; he ran the Instituto di storia dell'architettura. It was Tafuri who shaped the community of young scholars—Francesco Dal Có, Marco de Michaelis, Mario Manieri-Elia, Giorgio Ciucci, and Massimo Cacciari—who became identified with the Venice School.[5] Appointed by Tafuri as professor of aesthetics, Cacciari has often been acknowledged as an important intellectual force behind the Venice group, and as a major influence on Tafuri himself.[6]

Although Cacciari's contribution will be addressed, Tafuri's work—and the critical reactions to it—provide the focus for this chapter. From the late nineteen-sixties, the question of avant-gardist negation—and negation per se—occupied a central and self-conscious place in the arguments put forward by these historians and theorists. Particular attention will be given to Tafuri's account of the avant-garde and the articulation of negation in his short and most famous book, *Architecture and Utopia: Design and Capitalist Development* of 1973 (based on an influential 1969 essay), and the elaboration of these ideas in *Modern Architecture* (1976) and *The Sphere and the Labyrinth* (1980).[7] Koolhaas's intervention at Euralille explicitly nods at a key motif and social metaphor from Tafuri's work: the labyrinthine *Carceri* by Giovanni Battista Piranesi (see figure 2.2).

Tafuri paid close attention to the social-democratic and communist architectural and artistic projects of the twenties and thirties when this history was still scarcely engaged with by Western scholars. From Tafuri's arrival at the institute, innovative research on the Soviet avant-garde, spanning from the NEP to the Second Five-Year Plan, took place, which resulted in an international seminar in 1970; the associated publication included essays on the experience of those architects (such as Ernst May, Hans Schmidt, André Lurçat) who had migrated from Europe to work on major projects in the U.S.S.R.[8] In addition to feeding into Tafuri's own books, the research produced more interpretative essays and edited translations of architects' statements by other members of the team. From 1969, with the Soviet project still in progress, Tafuri and Francesco Dal Có, along with Giorgio Ciucci and Mario Manieri-Elia, also worked on a study of American urbanism covering the period between the Civil War and the New Deal (published in 1973).[9] The two-volume *Modern Architecture* (1976) resulted from another project between Tafuri and Dal Có initiated in 1971. Although they do not always command the critical status achieved by *Architecture and Utopia* and *The Sphere and the Labyrinth*, the value of this work is indicated by the fact that the publications resulting from these joint projects are still recognized as defining contributions. Collaborative work clearly appealed to Tafuri: he spoke of his late work on Mannerism in such terms, but in the late sixties and early seventies such projects were at their most vibrant. The energetic pace of this work and its desire to stake out critical parameters for a discipline is still palpable, and even where texts are individually penned, a genuine ethos of a shared research endeavor pervades the writing.

In many ways, Tafuri needs to be seen within the currents of Italy's communist culture and as a beneficiary of the expansive, and

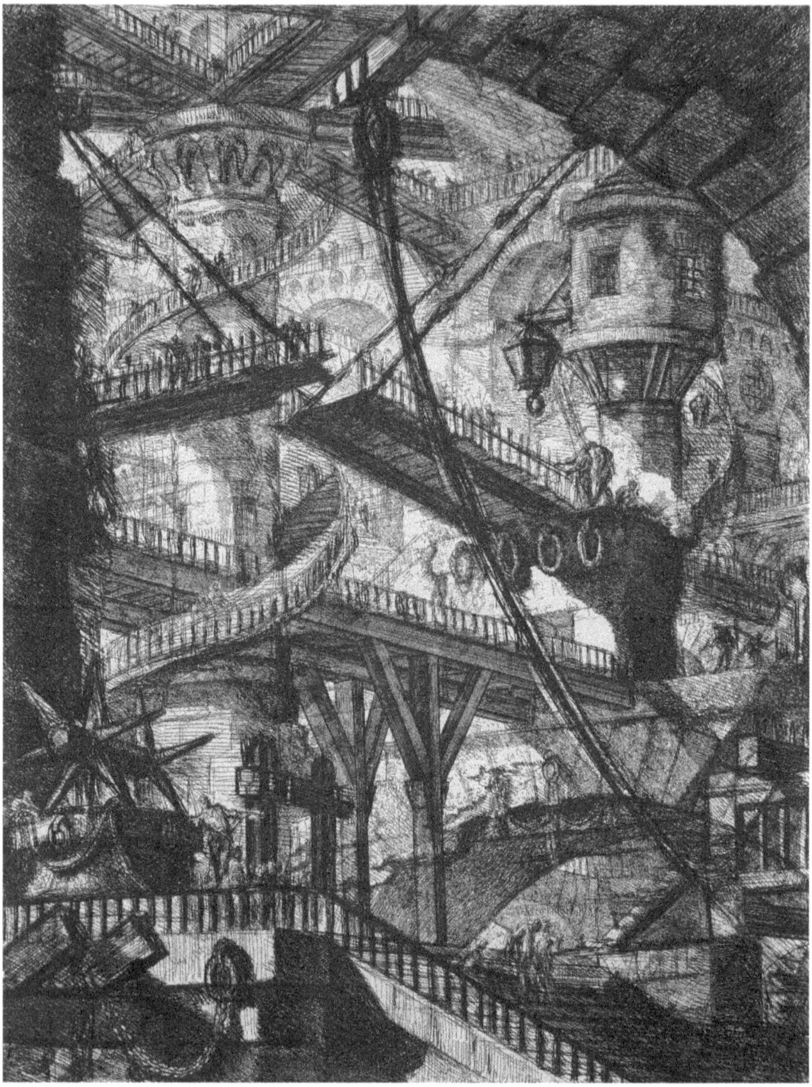

2.2 Giovanni Battista Piranesi (1720–1778), *Carceri VII*, 1760, etching. On Loan to the Hamburg Kunsthalle, Hamburg, Germany. Photo: The Bridgeman Art Library.

relatively open, intellectual atmosphere in and around the PCI, especially under Togliatti. Shaped through an engagement with the Marxist cultural analyses of Galvano della Volpe and Giulio Carlo Argan, and alongside the work of younger thinkers such as Alberto Asor Rosa and Franco Fortini, Tafuri's work, like theirs, is characteristically purposeful and exploratory, afraid neither of making forceful judgments nor of addressing complex social or aesthetic questions. His publications cover a wide array of topics, from his well-known work on the interwar avant-gardes, and the emergence—and fate—of architectural modernism, to his later publications on postwar Italian architecture and his exploration of the buildings of Renaissance Venice.[10] Methodologically, his research takes in both the large sweeps of grand historical analysis and the detail of microhistorical investigation, pursued either way with a rare intellectual ambition. Indeed, the reading of individual projects (and their particular crystallization of social tensions and architectural problematics) can be brilliantly inventive, something that is easily occluded when, in an effort to capture the ouevre's larger movements, commentators attempt to read Tafuri synoptically. Widely regarded as difficult, Tafuri's range of intellectual interests combines with a highly figured—and sometimes allusive—style of writing, characterized by a deeply embedded dialogism and by forms of dialectical logic.

Despite providing one of the major (and one of the earliest) critical appraisals of the avant-garde, only rarely can Tafuri's arguments be found alongside other key theoretical approaches on modernity, and his impact has been limited to the realm of architectural and urban studies. Although very different in mode, Tafuri's writing rates alongside Peter Bürger's *Theory of the Avant-Garde* for its critical acuity and its engagement with the prominent theories of modern capitalism (from Marx and Lukács, to Weber and Mannheim, Simmel, Benjamin, and Adorno). It seems that the combined barriers of the Italian language, poor translations, the specialism of architecture, and Tafurian difficulty have hindered his passage into the pantheon of theorists of modernity. Were Tafuri's work to belong to literature or even art history, it would certainly have a higher profile, especially if it were easier to distill its multiple interests into a packagable theory. Nevertheless, Tafuri's work has attained a high, if controversial, profile in architectural studies, enough so that Daniel Sherer can refer to "an on-going *discorso Tafuriano*."[11] The books *Architecture and Utopia* and *Modern Architecture* were translated in 1976. Between 1974 and 1979, five of Tafuri's essays were published by *Oppositions*, a theoretically orientated journal of architecture based in New York. (Dal Có saw five of his articles appear in the journal from 1978 until its demise in 1984, and Ciucci and Cacciari were also translated for it pages.) Despite a series of critical reviews in English-language journals, it took some time for this Tafurian discourse to gain ground. Even by the mideighties, when the participants from the Revisions study group at the Institute of Architecture and Urban Studies in New York published the proceedings of their "Architecture and Ideology" symposium in 1982, they still thought that negligible attention had been given in American architectural theory to the relations between architecture and politics. Nevertheless, through their symposium and its associated publication, *Architecture, Criticism, Ideology* (1985), which focused on Tafuri's work, the activities of the study group played a key role in the establishment of Tafuri for the English-speaking academy.[12]

The early eighties mark an important period for architectural theory. As Mary McLeod reflected in her introduction to the volume, the members of Revisions had doubts about what they saw as the new postmodernism's regressive nature: its formal and stylistic nostalgia and its association with the rising political conservatism of the time.[13] (As we will see in a later chapter, a similar concern will be echoed in Hal Foster's efforts to distinguish conservative and radical versions of postmodernism.) The group identified Italy as the beacon of an emergent "historical, materialist criticism" in architecture, which they contrasted to the poor state of American architectural thought; they even saw the beginnings of a Marxist architectural analysis to compare to Lukács's work on the novel or Adorno's study of music.[14] Whether positioned as a major part the Revisions group's projected materialist reclamation, or by *Oppositions* as the historical moment in architectural theory's "dialectic of autonomy and historical determinacy," it is clear that Tafuri's work was not simply experienced as a mere addition to the corpus of architectural history, but that a number of intellectual and political investments were being ciphered through it.[15] Reflecting back on the *Oppositions* project, and by way of a contrast with the formalism of Colin Rowe, K. Michael Hays positioned Tafuri as a worldly, historical, fallen angel.[16] Fredric Jameson attempted to pin down a figure who seemed to defy the categories of modernism and postmodernism, concluding that Tafuri's work should be seen as both antimodernist and antipostmodernist, or to put it another way, as representing the "neither . . . nor" of architectural theory.[17]

Tafuri's Dialectic of the Avant-Garde

Until the nineteen-eighties, Tafuri was preoccupied with understanding the work of the interwar avant-garde and the historical lessons of this project for his own generation of architectural practitioners and political activists. As we have seen, he traced the work of architects associated with the avant-garde as they encountered the new demands presented by both the Soviet and American contexts. His critical examination of, and skeptical take on, the "art into life" project associated with the avant-gardes was unusual at the time. In the late sixties and early seventies, interest in avant-garde culture was still largely concerned with historical recovery; this was particularly the case with those central and eastern European practices, which were still relatively unknown to Western scholars, but which were central to Tafuri's arguments. Working in Italy, however, Tafuri and his colleagues had available a wave of relevant publications and translations that had appeared in the midsixties, books addressing Russian Formalism, avant-garde theatre, and Soviet architecture.

Nevertheless, Tafuri's work was also unusual for its argument. In *Architecture and Utopia*, he challenges the ideology of avant-gardism, regards the appropriation of avant-gardist rhetoric by radicals with suspicion, and questions the avant-garde's own emancipatory claims. In this account, avant-gardist negation is not to be seen as a once-radical thing that had unfortunately been appropriated by capitalism's commercial and political machinery—not, to use Abigail Solomon-Godeau's memorable phrases, as "armed vision disarmed" or "weapon" become "style." Tafuri summarizes his arguments in his note to the second Italian edition of *Teorie e storia dell'architettura*:

> *The use of negation*: this new value, or, better, this new operating technique, has a constant function in the transformation of the capitalist-bourgeois crisis into models of development. From 1920 to about 1935, architecture was in the forefront in the battles of the dialectical conversion from Negative to Positive. Its crisis only comes at the precise moment in which, facing the reality of the *Plan*, the role of foreseeing or ideologically mediating the Plan ceases to exist.[18]

We will be returning to the key terms outlined in this passage. Unlike so many critical intellectuals working in Britain and the United States about a decade or so later, Tafuri's work is no retrieval exercise salvaging the radical strains of modernism. For him, the negativity of the avant-garde should be seen, from the outset, as wrapped up with capitalism's modern coming-to-being, its artistic innovations ultimately playing a role in social restructuring. Fredric Jameson captures something of Tafuri's argument, which, he describes as being:

> a powerful indictment of . . . "protopolitical" impulses in high modernism (the "Utopian" substitution of cultural politics for politics proper, the vocation to transform the world by transforming its forms, space, or language). Tafuri is, however, no less harsh in his anatomy of the negative, demystifying, "critical" vocation of the various modernisms . . . whereby the instrumentalizing and desacralizing tendencies of capital itself are ultimately realized through just such demolition work by the thinkers and artists of the modern movement. Their "anticapitalism" therefore ends up laying the basis for the "total" bureaucratic organization and control of late capitalism.[19]

Jameson holds on to some of the complexity in Tafuri's work; others are less subtle. In the hands of many subsequent commentators, Tafuri's argument is reduced to the claim that the avant-garde represents the research and development arm of capitalist modernization. In this chapter, such simplifications of Tafuri's account will be questioned.

Despite his trenchant criticism of avant-gardist negation, negation was nevertheless at the core of Tafuri's method. His approach draws on Cacciari's research on central European culture of the early twentieth century, particularly that of Vienna, and on Cacciari's philosophical interrogation of "negative thought" from Schopenhauer and Nietzsche—all conducted from within the framework of a political project of radical emancipation.[20] For some, this was a controversial, not to say paradoxical, argument. This trajectory, which at its far end engages the materials of irrationalism, is the product of thought conscious of both negativity's dialectic of reason and unreason and its place in the avant-garde heritage. In an important way, Tafuri and Cacciari can be seen as both Lukácsian and anti-Lukácsian, imbibing of those thinkers pilloried in *The Destruction of Reason* while choosing a conceptual menu from *History and Class Consciousness.* Tafuri's critique of the avant-garde adopts an analytic framework familiar to the Marxist intellectual tradition of the midtwentieth century, but does so in the service of some radically unconventional judgments.

K. Michael Hays suggests that Tafuri's work represents a form of historical determinism; similarly, Tomas Llorens describes Tafuri's work as "tracing architectural conceptions to some basic kernel in the economic domain."[21] However, Tafuri's method does not simply track the influence of the base on its architectural superstructure, but explores contradictions and limitations faced by the practice within its politico-economic context, and how architecture and urbanism become the means for "the direct transformation and utilization of ideology and culture as a technique."[22] As with the tradition of Critical Theory, Tafuri fuses an economic and social analysis with some of the central ideas of non-Marxist sociology—particularly Max Weber's. Lukács was again a key influence. There are some strong continuities here with Argan's work: the attention to the concept of ideology; the sense that progressive art partakes of "continual transformation, determination, and renovation of values"; the argument that, in its inability to see through the revolutionary potential of the interwar years in particular and the Enlightenment project in general, modern architecture had reached an historical impasse; his rejection of fantasies in organic, or natural, society; the belief that an analysis of culture's historical determinacy must be supplemented by political evaluation.[23] The Tafurian account not only draws on classic Marxist analysis; it is equally driven by the Weberian thesis of "disenchantment" and Nietzsche's account of "devaluation." The mix he achieves, however, has its own unique flavor.

Tafuri believes that the role of intellectuals and artists underwent a fundamental transformation during the twentieth century. Intellectuals no longer perform their traditional role as legitimators of values and ideologies; rather, they increasingly have come to play a modernizing role, in which the destruction of tradition and value is central to their remit. "In order to survive," Tafuri writes, "ideology had to negate itself as such, break its own crystallized forms, and throw itself entirely into the 'construction of the future.'"[24] No longer unproductive actors on the sidelines of the economy, intellectuals have gained a role within the realization of capitalist development as key members of a new technocratic fraction.[25] Tafuri cites as evidence for this shift a growing collaboration between the representatives of big capital and the emergent design profession (as, for example, in the case of the Werkbund). According to him, the avant-garde helped to acclimatize the public to the disruptions of the urban world. "Provocation," Tafuri writes, "was only the outer skin of a process in which the bourgeoisie took possession of the technological universe."[26] Transforming the experience of shock into the language of design, the avant-garde normalized negation's social ruptures through its formal and spatial tactics. Whatever the subjective intentions of individual avant-garde producers, Tafuri argues, their *objective* role in the process (however marginal) was to function as a force for modernization, thereby contributing to capitalism's changing requirements.

The Metropolis

The concept of the Metropolis—"the postulate of the intrinsic negativeness of the large city," as Tafuri puts it—played a major part in his thought and was central to his understanding of the interwar avant-gardes.[27] As he writes in *Modern Architecture*, the Metropolis is the very foundation of the avant-garde and "the real proving ground for all its proposals."[28] In his essay "The Metropolis and Mental Life" of 1903, Georg Simmel considers "the adaptations made by the personality in its adjustment to the forces that lay outside of it."[29] Modern urban experiences of constant speed, innovation, and change, he argues, produce "the intensification of emotional life due to the swift and continuous shift of external and internal stimuli."[30] Mental life takes on an "essentially intellectualistic character" as a necessary response to—"a protective organ" against—"the fluctuations and discontinuities" that threaten it.[31] As Simmel puts it:

> [The] money economy and the domination of the intellect stand in the closest relationship to one another. They have in common a purely matter-of-fact attitude in the treatment of persons and things in which a formal justice is often combined with an unrelenting hardness.[32]

Nerves and emotions are dulled through an overexposure to and overload of stimuli, producing a new form of behavior defined by the "incapacity to react to new stimulations with the required amount of energy."[33] The blasé type who emerges under these conditions is characterized by an attitude in which:

> the meaning and the value of the distinctions between things, and therewith of the things themselves, are experienced as meaningless. . . . money takes the place of all the manifoldness of things. . . . it hollows out the core of things. . . . They all float with the same specific gravity.[34]

This "devaluing [of] the entire objective world" also drags "the personality downward into a feeling of its own valuelessness," affecting the entire pattern of social interaction, which becomes characterized by "indifference," "reserve," "aversion," and even "hatred and conflict."[35]

Benjamin famously picked up on Simmel's themes and, in the Italian context, Cacciari—a founding editor of *Angelus Novus* and an early reader of Benjamin—developed the concept of the Metropolis in his essay "The Dialectics of the Negative and the Metropolis," prepared in the late sixties and published in 1973. Negation is the central category for Cacciari's analysis of German sociological thought. He writes:

> When Geist abandons the simple and direct relations of production, it no longer creates the city but the Metropolis. It is the Geist, not the individual, that *of necessity* inhabits the Metropolis. This is the objective reason for the Metropolis.[36]

In a move echoing Lukács's own development of reification, Cacciari extends Simmel's discussion, inflects it with Marx's analysis of the commodity, and establishes "the Metropolis" as the figure for the "life" of capitalism—its life as the general form for the rationalization of social relations. This rationalization, Cacciari writes, is "a process that abstracts from the personal and rebuilds upon subjectivity as calculation, reason, interest," extending beyond the experiences of working life to invade the most intimate pores of daily and psychic existence.[37] Retaining the German noun—as he does with *Verstand* and *Nervenleben*—Cacciari calls this sublimation of subjective by objective processes "*Vergeistigung*," or, to put it another way, the realization and becoming of *Geist*'s abstract spirit.

Vergeistigung here figures the process of reification or increasing social abstraction in modern life. The Metropolis, Cacciari argues:

> dissolves individuality into the current of impressions and reintegrates these, precisely by virtue of their constitution, into the overall process of Vergeistigung. In its first stage of evolution, the Metropolis uproots individuality from its conservative fixity; the process begun by this uprooting will of necessity lead to the dialectical reasoning that governs, measures, and directs social relations, the interest (*inter-esse*) of the Metropolis.[38]

With the destruction of the city—that is, the city conceived as an organic entity, or *polis*—it appears henceforth only as a lost ideal or "Goethian city," symbolizing a nostalgia for totality, plenitude of experience, and integrity of values. Thus, in the Metropolis, Cacciari maintains, while drawing out Simmel's monetary analogy in the light of Marx's opening chapter to *Capital*, everyone and everything become analogous to the market's universal equivalent—all concrete use-values, all qualitative meanings—and become subsumed under the sway of quantification and exchange. As he argues:

> the Nervenleben corresponds to the continuous and relentlessly innovated transubstantiation of exchange value into use value—that is, it corresponds to the necessary instance in which exchange value becomes real value. The intellect, the *Verstand*, in turn abstracts from the appearance of use value the substance of exchange value; it extracts *money* from the process and thus correctly reflects upon the commodity as such—that is, it once again produces merchandise.[39]

In Simmel's view, individuality and emotion take on the characteristics of exchange value and are defined by equivalence and quantification. The blasé type, Cacciari continues, is aware of this situation, knowing that he can buy commodities (both goods and people) but also that "he cannot get close to these goods, he cannot *name* them, he cannot *love* them."[40] Universal equivalence takes the form of spleen, responding to what Tafuri calls "the logic of assassination."[41] The Metropolis then is not only presented as a place of calculation and reason but also as the site of reason's "downfall."[42]

The Metropolis was not only a category employed to describe a process of social abstraction; it was also an abstract category. Cacciari uses the upper case "M" to register its heightened conceptual status. In this regard, the Metropolis has affinities with Debord's concept of the Spectacle, Lefebvre's exploration of everyday life, and Marcuse's account of one-dimensional consciousness. It goes without saying that all concepts are closely indebted to Marx's work on commodity fetishism and to Lukács's subsequent development of the category of reification. In each case these thinkers try to capture a complexity of societal processes, their contradictory manifestations, and their fetishistic petrification. Their categories attest to the conditions through which we live, to convey both the dislocation of reality and the reality of that dislocation, and attempt to follow the transmigration of conceptual qualities. Metropolis, like these other central categories, names capitalism at a certain stage of its development, along with capital's most widened social effects and impact on individual consciousness. These are highly generalized terms, subsuming—in ways that can be both a helpful shorthand and unhelpfully vague—a huge range of valences; yet, this abstraction also relies on specificity, hinging on the theory of the commodity (and, to an extent, the theory of value). This specificity is what makes "Spectacle" not "spectacle" in its everyday sense, and why the conditions invoked by Cacciari's Metropolis exceed what we usually understand to be the metropolitan condition. Spectacle and Metropolis are categories that concern social process and social relations. Yet, Spectacle and Metropolis carry the idea of rationalization much further than Lukács's defining account. "Spectacle" and "Metropolis" describe the abstractions of reified existence; they also serve as abstract categories; in addition, they run the risk of becoming reified categories as such.

The differences between these attempts to characterize modern capitalism are equally significant. Where Debord, for example, emphasizes the issue of representation, Metropolis turns on disenchantment and devaluation. (And just as Spectacle plays on ambiguity of the words "representation" and "image"—hence its deceptively smooth incorporation into theories of art and postwar cultures of the mass media—so Metropolis flexes the term "value," so that it simultaneously leans toward the axes of Marx-Lukács and Weber-Nietzsche). Although not reducible to the large city, Metropolis nevertheless concerns processes that are most often identified with urban development, or at least with the type of social relations ushered in with urban expansion, from Baudelaire's *flâneur* (as discussed by Walter Benjamin) to the contradictory flows of capital investment examined in modern geographical studies. Metropolis counters the idea that our built environment is an agglomeration of static objects—an urban sprawl of discreet buildings. But this argument is not just intended to shift our attention from buildings to people (this is not a move into phenomenological considerations of spatial and temporal embodiment). Rather, Metropolis is about seeing (so to speak) around, through and into all these apparently static entities (buildings, people, complexes, ideas), so that they might be understood as condensations of social process—as residues, crystallizations, or as petrified things—to see them, in other words, as if they were Benjamin's "dialectics at a standstill." Yet, these differences in content—between Metropolis and Spectacle—rest mostly on questions of emphasis and application; it would be easy enough to synthesize them into a general account of social reification. (Here, Debord's comments are prescient: "Without a doubt," he writes, "the critical concept of the *spectacle* is susceptible of being turned into just another empty formula of sociological-political rhetoric designed to explain and denounce everything *in the abstract*—so serving to buttress the spectacular system itself."[43]) Despite a considerable amount of shared political grounds between Cacciari and Debord, it is in the end their political-philosophical orientations—and especially the attitude toward nihilism—that most separate Metropolis from Spectacle.

Metropolitan Avant-Gardes

We need to keep in mind the profile of the Metropolis as we return to Tafuri's account of the European avant-gardes. Avant-gardists responded, he argues, not only at the level of their subject matter but also by registering the Metropolis at the level of form and the sign. In the work of Cézanne and the Analytic Cubists, Tafuri locates an emergent awareness of Metropolitan fragmentation, its disenchantment, and its dislocation of sign and meaning. But, he argues, the "loss of center" fully made its way into art with the Italian Futurists, who, placing themselves at the "head of the anonymous impulse sweeping away the ancient values" and the "total revolt of things and mobs," asserted an intellectual will-to-power.[44]

In their cabarets full of "bombarding words liberated of sense and syntax," Futurist art echoed the Metropolis.

> It was the machine that was now determining the modes of communication, and its messages were compounded of pure energy that had no need for syntactical nexuses. The technological language was based on something new: shock, pure signs assaulting the interlocutor all at once.[45]

As agents in the internal reshaping of capitalist social relations, avant-gardists' search for new forms, for new ways of making art or designing buildings, played an important role in sweeping away older modes of being. In exploring the relations and responses of Dada and De Stijl to the disenchantment of the Metropolis, Tafuri outlines his "dialectic of the avant-garde," a dialectic which opened with a classic contrast between Dada's "violent insertion" into the irrational and De Stijl's efforts to subject form to reason.[46] According to Tafuri, however, this opposition of rational form and irrational formlessness inverted. Dadaists immersed themselves "in the unformed," thereby displaying "the void that the end of values . . . leaves behind it."[47] Yet, it was by way of this very representation of chaos and nullification that Dada affirmed and sought control of modern reality. By giving form to Nietzsche's death of God, Tafuri argues, and by ironizing the anarchy of the Metropolis, Dada demonstrated its own constructive face, its own countertendency toward the control of that formlessness (see figure 2.3). Around the same time, in an effort to grasp the fundamental elements and language of form, De Stijl employed analytical procedures, but was, according to Tafuri, unable to recombine convincingly the fragments of this investigation. Instead of marking out an ideal space against chaos, De Stijl's "silent immersion in the structure of the city" was overrun by the experience of Metropolitan fragmentation.[48] Far from reason, Tafuri suggests, De Stijl could only produce "disarticulated recomposition" (see figure 2.4).[49] Its practitioners thus found that they had merely echoed the mechanical world, and that they had participated in Metropolitan decomposition. Consequently, these two avant-garde tendencies—which we might recognize as examples of Renato Poggioli's "negative" and "positive" avant-gardes—increasingly shared "a common language" of inorganic disintegration.[50]

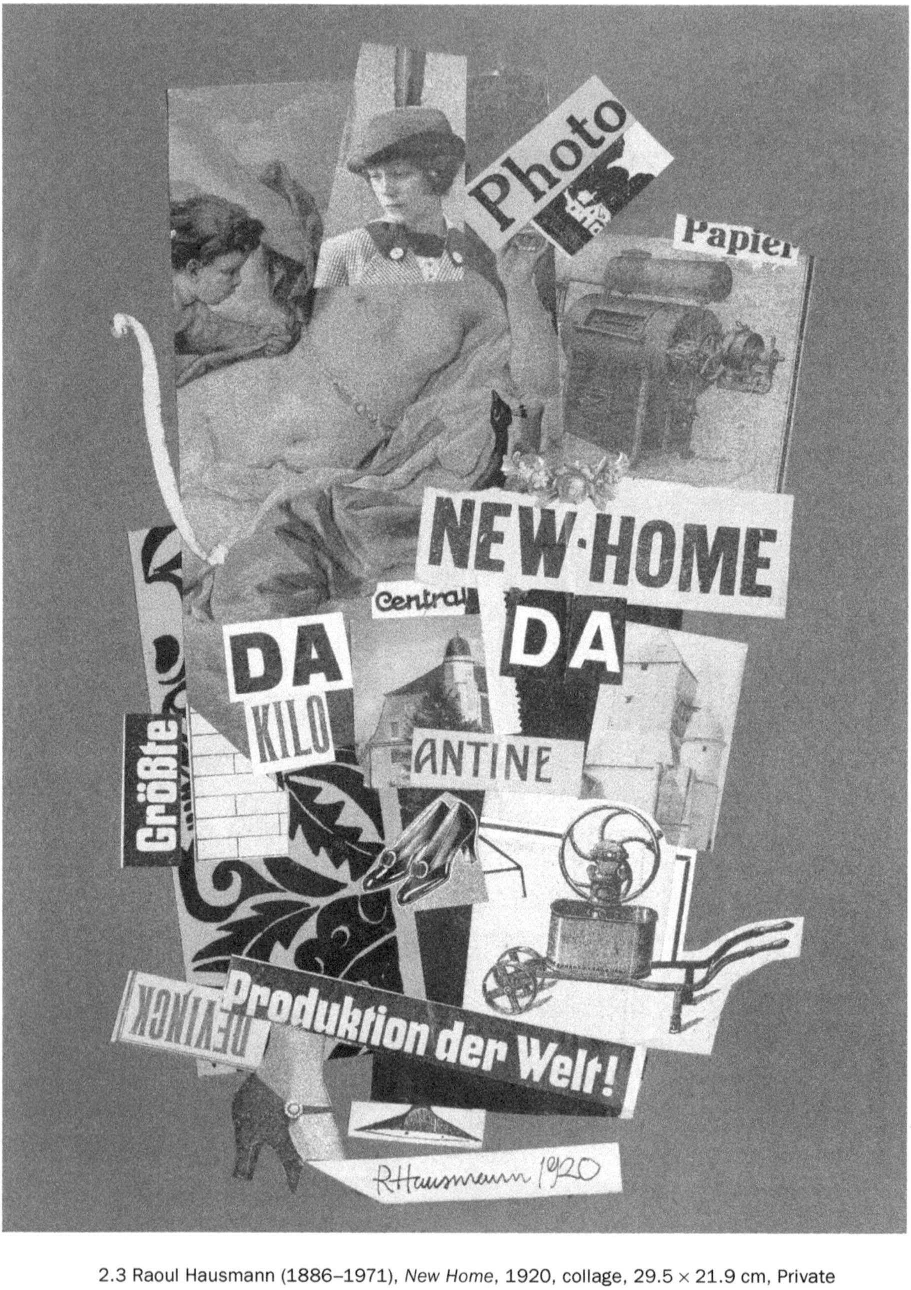

2.3 Raoul Hausmann (1886–1971), *New Home*, 1920, collage, 29.5 × 21.9 cm, Private collection. Photo: The Bridgeman Art Library. © ADAGP, Paris and DACS, London 2009.

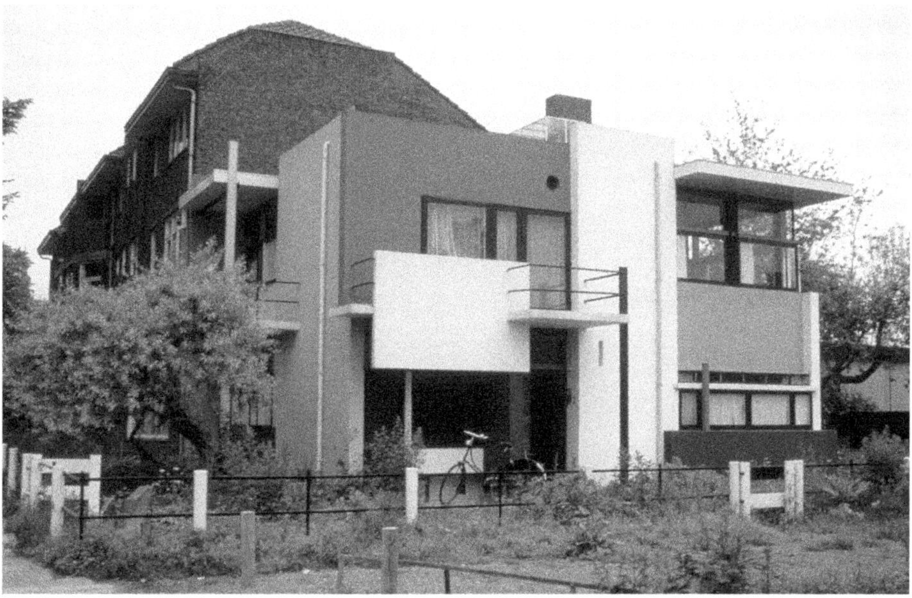

2.4 Gerrit Rietveld (1888–1964), Schroeder House, built in 1923–1924, Utrecht, Netherlands. Photo: The Bridgeman Art Library. © DACS 2009.

The year 1922 represents a major point of transition in Tafuri's account. At this point the distinct European avant-gardes—many prompted by the important exhibition of Russian avant-garde art at van Diemen's Gallery in Berlin—reoriented their concerns under the impact of Constructivism, creating a new international network. According to Tafuri, Dada "limited itself merely to specifying the instruments" and techniques adequate to the Metropolis—collage, assemblage, and montage—and "the mere enunciation of principles." The emergent Constructivist tendencies, in contrast, demonstrated how to put the new formal principles to "institutional use." "Constructivism was a metaphor for the technical organization of the real," he argues, and for "the dynamic articulation of signs that were completely disenchanted."[51] By seeking the sign's zero degree (as both prehistory and ahistory), the Constructivist avant-gardes, Tafuri insists, established building blocks based on the disenchanted sign (see figure 2.5). Evacuated of any qualitative value or given meaning, he argues, these signs could function as cultural analogues to universal equivalence. The disenchanted signs of Constructivism became available for constant reinvestment with new values, the

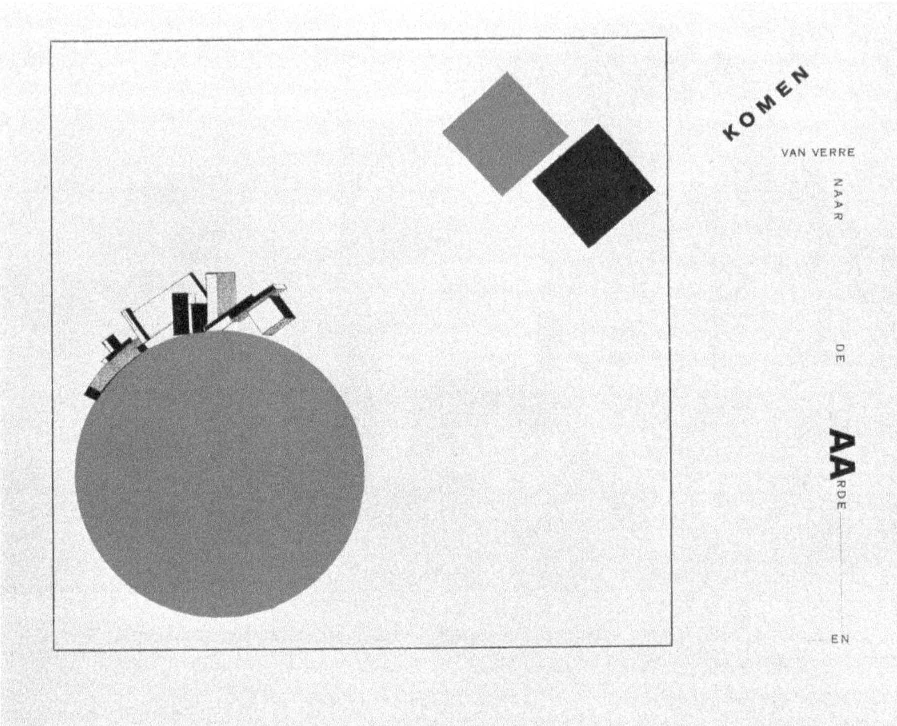

2.5 Eliezer (El) Markowich Lissitzky (1890–1941), Illustration to *The Story of Two Squares*, written by the artist, published in Berlin, 1922 (printed book), Haags Gemeentemuseum, The Hague, Netherlands. Photo: The Bridgeman Art Library. © DACS 2009.

logic of El Lissitzky's "Prouns" testifying to the fluid range of meanings possible: a child's fantastic storybook, the "ABC" and "Objekt" of European graphic modernity, an invocation for workers to return to their lathes, an appeal to fight in a civil war, even photographic displays for international exhibitions dedicated to the press or to hygiene. From *Architecture and Utopia* through *Modern Architecture* to *The Sphere and the Labyrinth,* Tafuri's argument on the practices of early-twentieth-century art remains remarkably consistent.

As Tafuri acknowledges, the means for arriving at the disenchanted sign were rarely straightforward, and in the process "mystical residues and disenchanted cynicism alternated and were superimposed."[52] Moreover, following Jan Mukarovsky, Tafuri understands Russian Formalism as aspiring to an end "beyond formalism." Viktor

Shklovsky's work is seen as a "laboratory" period in the endeavor "to reconcile the internal dynamics of the artistic structure with the context in which it is set."[53] The device of *ostranenie*, or defamiliarization, Tafuri argues, is best comprehended as a strategy of displacement, distortion, and deformation intended to retrieve words from "semantic exhaustion," to recover original meanings "for things and concepts," that is, as *The Resurrection of the Word*.[54] Avant-gardist techniques such as montage, juxtaposition, or dislocation, then, should be seen not as merely echoing Metropolitan fragmentation but as the opposite, as attempts to close the gap between sign and meaning. Thus, in the case of the Russian Formalists, Tafuri argues, their move to abstraction, far from being a "flight from the real" (as often perceived), results from the very "effort to rejoin reality" and to break free from reification.[55]

The effort to rejoin reality, however, proved to be the central problem not just for Shklovsky but for the entire avant-garde ambition. Indeed, Tafuri argues that "the very concept of the avant garde" contains a contradiction, one that manifests as "moments of rupture—as moments of conflict" emerging "along the problematic boundary line separating the avant-garde itself from the reality principle."[56] Although aspects of this argument can be found in his earlier writing, the point is particularly well developed by the early nineteen-eighties, in both the essays on Shklovsky and *The Sphere and the Labyrinth.* Tafuri grasps that the problem revolved around separation. It is important, however, to keep in view the extent to which this argument diverges from those accounts of the avant-garde rooted in the analysis of the Frankfurt School. The explicit attempts to bring together art and reality, to collapse the difference, do not pose the threat to "nonidentity" in the Tafurian argument; instead, the central dilemma facing such artistic projects is the problem of failure. There are two moments to this contradiction of the avant-garde: firstly, avant-gardists find that they are unable ever to affect their reality in the ways they intended; secondly, their strivings to fuse art with life turn into their opposite, reaffirming the very condition of alienation that many sought to transcend. In Tafuri's account, then, this troubled dialectic generates an ever-exacerbating tension for avant-garde projects.

At this point, it is worth observing a few of the examples that Tafuri gives of this tension. He regards Juan Gris's Cubism as blind to the achievements made in the "Analytic" phase of Braque and Picasso's work; this is a fairly conventional judgment, but Tafuri's reasoning was unusual. Tafuri believes that under the influence of Henri Bergson, Gris tried to reconcile subject and object, sign and meaning—the very disjunction that Georges Braque and Pablo Picasso correctly acknowledged as an irreconcilable condition. Gris's Cubism, in other words, attempted to mend the gap that had been opened in the Metropolitan situation. Yet, Tafuri asserts, Gris's project was doomed; his efforts to reconquer reality could "only fluctuate over it, imposing its own synthesis as supreme will-to-form."[57] What Tafuri seems to be arguing here is that Gris was attempting, voluntaristically, to repair the social rend; the strength of Picasso and Braque, in contrast, was that they grasped the new conditions of modern capitalism without attempting to provide any aesthetic balm. Although he is much more sympathetic to their projects than he was to that of Gris, Tafuri also viewed El Lissitzky's "Proun" paintings and Tatlin's counterreliefs as similarly "condemned to '*tend towards*'" the real, and to remain "pure aspiration, to live enclosed in a dream."[58] With his "Monument to the Third International," Tatlin hoped, Tafuri argues, that his historical moment—the communist revolution—could sidestep the void between conventional signs and their meaning, imagining that his tower could reconnect art and life precisely because its materials were "iron, glass, and revolution." To believe that art in the modern period could merge with reality was, for Tafuri, deeply misguided; form, he insists, could not overcome the problem of separation, it did not "have the power to 'link.'"[59] In "The Stage as 'Virtual City'" from *The Sphere and the Labyrinth*, Tafuri also develops his dialectic of the avant-garde. Employing theatre to recover "a portion of unalienated space," avant-gardists tended to posit an immediate identity between the theatrical sign and expression, body and space.[60] They wanted, to put it simply, to dissolve theatre qua theatre. These various radical theatrical projects to create counterrealities—from Adolphe Appia to Walter Gropius, Vsevolod Meyerhold, and Erwin Piscator—could only be pale alternatives to reality. They were, Tafuri notes, always "superfluous" to the Metropolis itself: "The real city is already total theatre." In a passage resonating with analogies, he explains that:

> Gropius's and Piscator's problem is how to return the theatrical "community" to the state of a chorus in the face of the tragedy of

> History; but the technological languages that have taken over the *Totaltheater* sing independently the hymn of victory of the negative that has taken over the real from which that theatre tries to isolate itself.[61]

Hugo Ball and Kurt Schwitters were more clearheaded, Tafuri believes, possessing a greater awareness of the Metropolis. The "victory of the negative"—of estrangement, alienation, and reification—conquered art through the very procedures that artists devised to counter its effects. Thus, Tafuri concludes, the "tension toward the real was valid only so long as it remained just that, so long as it did not pretend to arrive at concrete results."[62]

Ultimately, for Tafuri, the avant-garde depended on "intellectual work that stopped at the doors of industrial production, that touched the world of labor only tangentially, that did not abandon the field of pure ideology."[63] Between artistic form and the real there was a "swerve," as Tafuri puts it, adopting the force of the Shklovskian "knight's move." In trying "to break the barrier between the language of forms and that of existence," the avant-garde also tried to extinguish the veering "between the work and what is other to it, between the object and its conditions of existence, of production, of use."[64] This necessary distance and separation may appear to echo Adorno's anti-identity arguments and debates over autonomy, particularly as they manifested themselves in his aesthetic theory. Yet, Tafuri's metaphors—to swerve, to fluctuate over, to show tension toward—indicate a distinct mode of thought, one that tries—almost physically—to feel the processual dynamics within avant-gardist actions, to understand and empathize with their desire to effect their reality. Avant-gardist hubris, self-deception, and naivety are challenged in his analysis, but Tafuri's exploration of these contradictions is, for all the criticism, not unsympathetic in tone. The Tafurian dialectic of the avant-garde differs from ideas associated with the Frankfurt School. The necessity of nonidentity has another—nonphilosophical—dimension to it. Tafuri's work on the avant-garde must be seen as an allegory of the political question of reform or revolution, with attempts to modify capitalism for the better understood as being akin to the dilemmas faced by the avant-garde: rather than face the Metropolis, socialist reformism (including that advanced by many Marxists) attempted to repair society's fissures. Tafuri rejects the idea that aesthetic revolutions could achieve a resolution of the modern world's contradictions; he does not even conceive of social revolution in terms of synthesis or restitution. The concept of revolution is not, for him, some general panacea; rather it is simply the condition for beginning to work on reality in a nonillusory fashion. Commencing this task, he believes, must entail confronting the new Metropolitan situation in all its negative force: grasping its conditions, entering into, and working with and through, them. In this view, the fusion of art and life is only ever a "fiction" of "pure intellectual elaboration." There is, to invoke the terms of Bürger's *Theory of the Avant-Garde*, no sublation, "false" or otherwise, whether on art's grounds or not; and the threat of art's subsuming by life praxis does not particularly trouble Tafuri. More to the point, this fiction of art-life fusion, he argues, "*had to remain just that*: a horizon constantly shifted forward" so that it could act as a catalyst for action.[65]

Overcoming Anxiety

Tafuri understood the avant-garde as a complex and contradictory phenomenon, which responded to the Metropolis's chaos and fragmentation of experience sometimes defensively, sometimes by actively acknowledging the new situation. The crucial issue, according to Tafuri, was what attitude to take to this condition, more precisely:

> how to shake off the anxiety provoked by the loss of a center, by the solitude of the individual immersed in revolt, of how to convert that anxiety into action so as not to remain forever dumb in the face of it.[66]

In his account of the avant-garde's responses—understood now as a political or philosophical view—we begin to see the elements of a strategy for engaging with the Metropolis that is advocated by Tafuri, the moments in which the description and analysis of the avant-garde shifts into a politics. Alluding to Vladimir Mayakovsky's work of 1913, Tafuri describe the "revolt of the objects" as "the dominating motif of bourgeois anguish."[67] But while Russian Futurism sought immersion "in the sea of disordered objects to attempt the last possible synthesis with them," it was the Italian Futurists who broke with "all vindication of lost objects."[68] Anxiety over the Metropolis had to be supplanted by embrace. "It was necessary," he argues:

> to pass from Munch's *Scream* to El Lissitzky's *Story of Two Squares*: from the anguished discovery of the nullification of values, to the use of a language of pure signs, perceptible by a mass that had completely absorbed the universe without quality of the money economy.[69]

The Metropolis was not to be—could not be—evaded but had to be engaged productively. The new language emerging from the Metropolis—the possibilities presented by the breaking up of syntactical connections and the disenchanted sign—could only be unleashed once the avant-garde had "neutralized the paralyzing anguish that can only contemplate itself."[70] The fear of the present conditions and nostalgia for an imagined older social order could only end in a disconnected solipsism and was hopelessly unrealistic. For Tafuri, the way ahead involved actively embracing the given situation.

The turning point in Tafuri's dialectic of the avant-garde paralleled historical transformations in the role of ideology and the character of capital accumulation, and was echoed in new forms of social and economic theory. The broad shift from laissez-faire to monopoly capitalism—and the particular consolidation that followed the 1929 crash on the New York stock market—provided the framework for these new ideas. A distinctive feature of Tafuri's account was his refusal to posit the crisis of the avant-garde on the cataclysmic effects of political dictatorship. Not that they were ignored; Tafuri's Marxism understood both Hitler's Germany and Stalin's U.S.S.R. within the emergent socioeconomic developments of the interwar period. If these political developments were acknowledged as dominant and dominating events, they did not represent absolute exceptions for Tafuri but were part of the same economic fabric as liberal capitalist states. As Tafuri emphasized, the histories of modern architecture's crisis were typically framed through the political impact of Stalin and Hitler; in his view, however, the problem should be located with the restructuring of capital and the realization of the modern economic form.[71]

In *Architecture and Utopia*, Tafuri introduces his discussion of this social transformation by describing a shift from "Utopia" to "Plan" (or

"Project"). This epochal change within capitalism is at the heart of Tafuri's argument that intellectuals and artists have taken on new social roles, ones involving a constant intervention into, and reorganization of, reality. As the ideology of stable values gave way, the avant-garde's negation of tradition unleashed action from the strictures of ethical justification. As Tafuri puts it, "Liberation *from* value in this sense signifies establishing the premises for action in *that* reality, in that field of indeterminant, fluid, and ambiguous forces."[72] The very idea of utopia becomes transformed from an ideology of anticipated ideal and into a real working concept. Similarly, instead of being understood as a historicist repository of values, history comes to be seen as an event.[73] Ceasing to be ideals counterposed to the anarchy of capitalism, form and order are refigured as operative moments within, and for, it.[74] Tafuri points to parallels in economic science, such as in the work of John Maynard Keynes, which instead of seeking ways to stabilize economic conditions worked with conflict and contradiction. Devising models "*starting from the crisis*" rather than setting its theories "abstractly *against* it," chaos was conceived as something to manage, and economic crises seen as motors working for capital. Or, as Tafuri describes it, the new economic theory aimed to work with "the negative . . . inherent in the system."[75]

We will be returning to these arguments, but it is important to note that, for Tafuri (and contrary to the everyday sense conveyed by the word), "plan" signifies the very opposite of a fixed and rigid model, indicating processes of constant intervention and revision and aiming to adapt to capitalism's contradictions at ever-higher levels. This meant that negativity—the transitory, the temporary, the contingent, and the oppositional—is understood as being central to social and economic development, the plan harnessing this force *as* capital's power. By clearing the way for active intervention in the here and now, this antihistoricism allowed the avant-gardes "to explode towards the future," enabling them to find a role within the emerging "planner-states" of the interwar years.[76]

> For the avant-garde movements the destruction of values offered a wholly new type of rationality, which was capable of coming *face to face with the negative*, in order to *make the negative itself the release valve* of an unlimited potential for development.[77]

We have seen, then, that the avant-gardes participated in this new climate of dynamic intervention "once the environment was reduced to things devoid of significance," when art's syntax had been reduced to pure geometrical signs and when its function shifted from symbolism to one of perceptual reorganization.[78]

Of course, some members of the avant-garde attempted to do more than just flirt with the "reality principle"; nevertheless, in Tafuri's view, they could not evade the founding contradiction of avant-gardism. As the avant-garde moved into design—and more specifically into urban design—the "crisis of the object" unraveled architecture itself. The conventional architectural object (the individual building project) denied the reality of the Metropolis by attempting to turn its back upon capitalism's urban flux. Even as extensive a project as Karl Ehn's Karl-Marx-Hof in Vienna (1927–1930)—a superblock estate on the scale of a small town, providing services for its 5000 inhabitants—only replicated this same problem on a larger scale (see figure 2.6). From the sixties, the American superskyscraper—Tafuri's examples are the John Hancock Building in Chicago and the World Trade Center in New York—broke away from the disenchanted relations that, he believes, American cities had previously fostered. An "event" on the skyline, a "disenchanted mountain" in a "city without qualities," the classic skyscraper was "intent on dominating . . . the unnatural forest of the

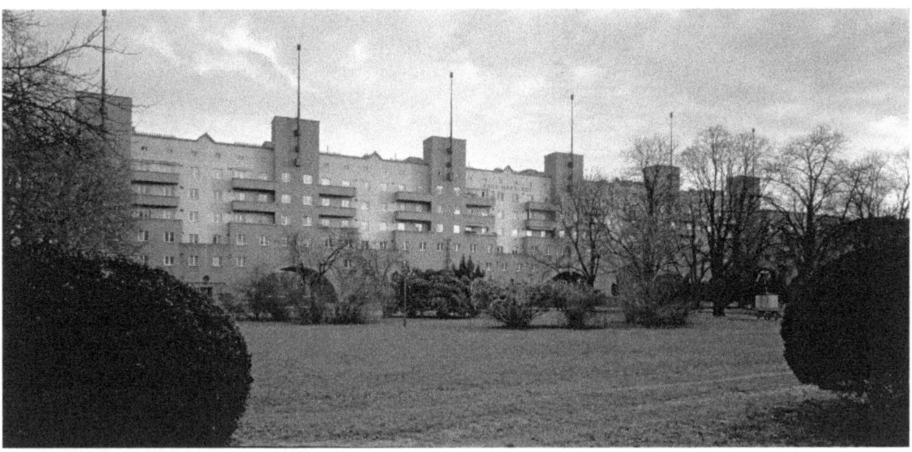

2.6 Karl Ehn (1884–1957), Karl-Marx-Hof, built in 1927–1930, Heiligenstädter Strasse, Vienna, Austria. Photo: The Bridgeman Art Library. © Courtesy of Paul Maeyaert.

metropolis."[79] In contrast, these superskyscrapers had each become a self-contained, "gigantic antiurban machine." The political associations involved are apparent:

> It is not emerging urban masses that erupt on the skylines of Manhattan, Chicago, and other American metropolises but, rather, antiurban paradoxes, artificial technological "miracles." The insertion of such structures into the two-dimensional grid of the city tends to negate the city itself in a desperate effort to escape its irrationalities.[80]

When conceived as an isolated object, or as a self-contained structure, architecture in Tafuri's view could be no more than an "ambiguous object" in the "Metropolitan 'Merz.'" Some designers attended to the urban process itself and understood the individual building as an entity absorbed within the city. Once again, however, this move fails to address the fundamental problem because so many urban planners imagined that they could reimpose form and rationality over the totality of the Metropolitan environment. And so, according to Tafuri:

> the entire cycle of modern architecture and of the new systems of visual communication came into being, developed, and entered into crisis as an enormous attempt—the last to be made by the great bourgeois artistic culture—to resolve, on the always more outdated level of ideology, the imbalances, contradictions, and retardations characteristic of the capitalist reorganization of the world market and productive development.[81]

The prevailing conditions required "architecture," in its classic sense, to self-negate and required that urban design fully acknowledge the contradictory processes of capital accumulation. The "environment" with which the planners now worked was amenable neither to formal experimentation nor to grand social syntheses; conceiving the metropolis as one gigantic village, or as an agglomeration of villages, was not to have begun to conceive, let alone understand critically, socially, politically, and philosophically, the "Metropolis." Moreover, for Tafuri, once the avant-garde came "within the sphere of the reorganization of production in general," they found themselves to be "the objects and not the subjects of the Plan."[82] In coming "face to face" with the negative, avant-gardists were increasingly forced to abandon tragic consciousness and to confront, in other words, reality in all its contradictions and reified confusions, its "masks," its "illusions," its invisible processes, its being "beyond synthesis."

Tafuri's hostility to tragic consciousness obliged him to pursue the logics of devaluation and disenchantment to their extremes. As he points out in the preface to *Architecture and Utopia*, we must accept that we are left with "form without utopia," with a "return to pure architecture."[83] Once faced with the crisis of the object, architecture becomes "a spectre of itself," a condition best exemplified by architectural contributions to the artificial environments of the university campus or the international exhibition: the "cities" of formal experimentation.[84] Buildings such as Mies's "Barcelona Pavilion," Tafuri suggests, cease to resist and give up trying to repair the damage wrought by the Metropolis. As he puts it some years later, "what is left to pursue is empty architecture."[85] In *The Sphere and the Labyrinth*, this argument is cast in metaphysical terms; he speaks of "the limit that separates language from silence."[86] Tafuri favors "silence"—the "absolutely asemantic quality"—rather than "noise," as the most appropriate response to the world "without quality."[87] Mies's late buildings, he argues, are "objects that 'exist by means of their own death,' only in this way saving themselves from certain failure."[88] For Tafuri, then, it is in works such as these that the "'negative' of the metropolis" is faced. Accepting its elements as "pure signs," this architecture abandons any attempt to reclaim, reconstruct, or reimpose meaning; having understood contemporary negativity, this is architecture "conscious of the impossibility of restoring 'synthesis.'"[89]

Empty architecture, architecture that is no more than a "spectre of itself," the inescapability of Metropolis's *Vergeistigung*, "form without utopia," avant-gardes that became the "objects of the Plan," and even approving nods toward Schopenhauerian abandonment of all will and representation: given all this, it should not be surprising that Tafuri's work has been taken to be a pessimistic assessment of art and the prospects for social transformation. The accusation of Tafurian pessimism is made persistently; even some of Tafuri's most astute readers resort to this claim. T. J. Clark—himself once on the receiving end of a similar charge from Michael Fried—refers to Tafuri (along with Foucault) as a "rank pessimist."[90] Fredric Jameson returns to Tafuri's arguments on a number of occasions. In an essay that, while deeply flawed, remains one of the best on Tafuri, his work is compared to Adorno's negative dialectic and Roland Barthes's zero-degree writing. Yet, despite noting that the terms "optimism" or "pessimism" are inappropriate in assessing Tafuri's position—"silly" and "frivolous" are the words he uses—Jameson concludes his essay by designating Tafuri an example of "cultural pessimism" and counters with the "positive" alternative of "Gramscian architecture" and "enclave theory."[91] In a subsequent article, Jameson describes Tafuri's argument as "perhaps the bleakest of all and the most implacably negative."[92] One might presume that Clark and Jameson are not making an everyday charge of pessimism. Neither thinker would be particularly distressed by Tafuri's antireformist perspective or by his relentless dismissal of illusions in capitalism's potential beneficence, let alone his capacity for socioeconomic analysis or his critique of the mores of architecture's professional ideology. Tafuri, of course, represents the most sustained and serious challenge to the critical utopianism that interests Jameson.[93] The pessimism charge continues to be cause for comment well into the nineties, having clearly taken root as an unwritten assumption in the *discorso Tafuriano*. Some try to distance themselves from it. While not elaborating their point, Hélène Lipstadt and Harvey Mendelsohn remark that, while understandable, such an accusation is nevertheless based on a "superficial reading."[94] K. Michael Hays, obviously responding to Jameson, tries to justify Tafuri's bleakness as, in part, due to the conditions of the economic and political crises of the nineteen-seventies.[95] Yet, Tafuri himself refers to this pessimistic reading of his arguments in his preface to the English edition of *Architecture and Utopia* in 1975—indeed, he alludes back to the reception of his writing in the sixties. In the midst of the world oil crisis and the social turmoil in Italy—not to mention the traumas undoubtedly being wrought by the PCI's historic compromise[96]—Tafuri explicitly and trenchantly distances himself from the accusation of being doom-laden. I take his repudiation of pessimism seriously.

Part II

"What is Bazarov?" Arkady smiled. "Would you like me to tell you, uncle, what he is exactly?"

"Please do nephew."

"He is a nihilist!"

"A what?" asked Nikolai Petrovich, while his brother lifted his knife in the air with a small piece of butter on the tip and remained motionless.

"He is a nihilist," repeated Arkady.

"A nihilist," said Nikolai Petrovich. "That comes from the Latin *nihil—nothing,* I imagine; the term must signify a man who . . . who recognizes nothing?"

"Say—who respects nothing," put in Pavel Petrovich, and set to work with the butter again.

"Who looks at everything critically," observed Arkady.

"Isn't that exactly the same thing?" asked Pavel Petrovich.

"No it's not the same thing. A nihilist is a person who does not take any principles for granted, however much that principle may be revered." . . . .

"Yes. It used to be Hegelians, and now there are nihilists"

—Ivan Turgenev[97]

Negative Thought

As we encounter the arguments about negative thought, Tafuri's alleged pessimism seems only to deepen. As he notes, "The avant-garde, brought back to its elementary principles, was . . . obliged to reveal its cards completely, to recognize its own origins in 'negative thought.' "[98] In his discussion of negative thought, as with the Metropolis, Tafuri again draws on ideas developed by Cacciari from the latter's engagement with Baudelaire, Nietzsche, Weber, Simmel, Keynes, and Benjamin. According to Cacciari, negative thought is the ideology most appropriate to the Metropolis; it represents "the discovery of the negativity of the Metropolis itself," that is, it "presupposes contradiction" and devaluation [*Entwertung*]; it recognizes that everything and everybody is engulfed in the Metropolis; it understands that "no aura can survive" and refuses the "prayer for consolation."[99] Indeed, Cacciari argues, the negative is negative "precisely because it is *Entwertung*." As "the motivating force of the age of technique," Tafuri writes, nihilism claims "for itself the field of differences between the project and utopia."[100] This statement is from the mideighties, but even as early as *Architecture and Utopia*, he can be found arguing that salvation lies not in revolt but in surrender, in the total submission to the dynamics of the Metropolis, in the acceptance of the void between sign and meaning—and in reveling in this situation.[101]

As he puts it elsewhere in discussions of Dada, one must acknowledge and indulge the Metropolitan situation, discover the "value of non-value," and *participate* in this condition. "To save oneself one must lose one's self," he ventriloquizes these artists' response, "one must resign oneself to being submerged in the chaos, one must make oneself sign among signs."[102] Discovering his own commercialization, the subject learns—and here Tafuri echoes Benjamin—to enjoy "being simultaneously torturer and tortured."[103] In a discussion of methodology in *The Sphere and the Labyrinth*, Tafuri suggests that historiography too entertains a fiction when it believes that it can fill the gap—the silence—between history and its object. Criticism, then, needs to throw itself into crisis along with the object; the self-reflexivity of historical work has to be relentless in order to remain conscious of its own murders.[104] Not only art and architecture, but also critical thought, must explore that negative space; it "descends into the interstices of techniques and languages" so as to elaborate the "residue" at the "insuppressible boundary" of silence.[105]

The issues raised by nihilism, negation, and the dialectic occupied a prominent place in the debates of the New Left, and the version of this argument elaborated by Cacciari and Tafuri—and they were by no means alone in this—developed along explicitly anti-Hegelian lines. Tafuri makes the point in *Architecture and Utopia*:

> "Negative thought" had enunciated its own project for survival in its refutation of the Hegelian dialectic and a recovery of the contradictions this had eliminated. "Positive thought" does nothing but overturn that negativeness on itself. The negative is revealed as such, even in its "ineluctability."[106]

Many key thinkers of the Left-Hegelian tradition resist the moment of sublation in the Hegelian dialectic, the *Aufhebung* often being seen as a closure on behalf of an unwarranted authority, or as a premature cessation of the impulse of freedom. From the Bauers in the mid-nineteenth century to Adorno's negative dialectic, such thinkers seek to release the negative from the strictures of a hegemonic power, whether politically or philosophically conceived. Elements of this Left-Hegelian argument are to be found in Tafuri's writing on architecture, but his antagonism to dialectical synthesis is focused through his opposition to searches for pre-Metropolitan wholeness and articulated through the trenchantly anti-Hegelian rhetoric common among parts of the New Left. Moreover, as we have seen in Tafuri's and Cacciari's accounts of the avant-garde, the negative is no longer simply allied to the emancipatory "force moment," its power being compromised by capital's appropriation of negation's dynamic. This preference for a form of anti-Hegelian negative dialectic can be seen not only in Tafuri's explicit discussions of negative thought and nihilism, but also in his wider approach to history; we have already encountered an example of this orientation in his distaste for the synthesizing tendencies of Juan Gris. Sharing the same large interpretive schema that can be found in the writing of someone like Giulio Carlo Argan—where the canonical history of art and architecture is renarrated dialectically through a play of contrasting movements, tendencies, or qualities—Tafuri nevertheless rejects its essentially progressive orientation.[107] Argan was not shy of finding aporetic moments in capitalism, but Tafuri pushes this sensibility still further so the dialectic (if it can still be called that) hovers at a breaking point. Tafuri argues, for example, that the "unity in variety" attributed to late Baroque architecture developed a conception of the city characterized as "order and chaos, regularity and irregularity, organic structure and the lack of organic structure."[108] Yet, in the Tafurian view, such contrasts do not add up (let alone sublate), this lack of coherence only deepening and escalating through history in a series of ever-exacerbating tensions and crises—history told, in other words, as aporia.

Referring to the work of Tafuri and Cacciari as "neo-avant-garde" and "radical gauchisme," Llorens, in one of the most probing critiques, argues that their method proceeds by means of paradoxes rather than dialectical contradictions. Their claim—that their "dialectics of negativity" was a Marxist anti-Hegelian approach—is, as far as Llorens is concerned, an example of pre-Kantian Nietzscheanism:

> The role of "negation" in Hegel's dialectics is often misunderstood. Hegel made it correspond to the second of Kant's "regulative principles of reason" . . . or, more properly, "determination." It can be said that Marx "inverted" Hegel by transferring this category from the epistemological to the ontological domain—or rather, by giving priority to the latter; in this sense he went further than Hegel in his critique of Kant's anti-metaphysical stance.

> But for this interpretation to make sense, it is necessary that the original framework, i.e., the postulation of a *transition* between the epistemological and the ontological, be retained. This is precisely what makes it possible to speak of marxist *dialectics*. The "dialectics of negativity" mentioned by Cacciari and Tafuri seem to refer rather to Nietzsche. Like Marx, Nietzsche "inverted" Hegel by postulating the priority of the ontological. But unlike Marx he did so by breaking it free from the critical framework of epistemology. Therefore there is no transfer, in Nietzsche, of logical categories, and his "negativity" is "negation" in a pre-kantian, material sense. How one can then speak of dialectics—Hegelian or not—I do not understand.[109]

Formally, Llorens may well be right, but, since Nietzscheanism is wielded so self-consciously and explicitly by Cacciari and Tafuri, his criticism hardly bites. Among the first of the "Nietzschean Marxists," Tafuri and Cacciari recognized the implications of their permanent suspension of sublation. There is rather more Nietzsche at work in their approach than in the thinking of many of their predecessors in the Western Marxism tradition (one thinks of Adorno or Lefebvre), and Tafuri is very early in adaptating this philosophy for cultural analysis. But—and it is important not to lose sight of this—their Nietzscheanism was conceived as being in the service of a radicalized Marxism.

It should be noted that even proponents of Left-Hegelian thought have found themselves open to the charge of nihilism. It haunts radical attempts to work with the dialectic, always accompanying the emphasis on its "pulse of freedom."[110] (Some even find nihilism centrally at work in the dialectical logic of Hegel himself.[111]) In effect, Llorens's critique rehearses key arguments from the history of twentieth-century thought: Adorno and Lukács, Adorno and Benjamin, Habermas and Foucault. Perhaps more tellingly, his own political allegiance blinds him to the ways Tafuri and Cacciari posed a radical critique of the orthodox Left, and thus Llorens fails to recognize, and account for, the methodological and political transfers at work in their thought. Adopting Benjamin's sado-masochistic image, Tafuri and Cacciari not only repeated one of the motifs from the *Trauerspiel*, they also took on a methodology that sought to transpose qualities from object to subject.

The procedures appropriate to—indeed, necessary for—the Metropolis were not only to be addressed within their objects of study (the avant-garde, architecture, or the genesis of negative thought), they were also required within their own approach to, and understanding of, those objects. Merely reflecting the Metropolis, Cacciari explains, "would be to reflect it not at all: between the forms and modes of such a simple reflection and the specifically dialectical structure of the Metropolis, no consistency is possible."[112] Characteristic of the Metropolis, such transitions from object to subject are also incorporated as a core principle of negative thought. Tafuri and Cacciari aim to capture both moments of this transposition—the external and the internal moments, so to speak. In this sense they repeat—and reverse—an experience of the avant-garde, who in their encounter with capital's negativity, we will recall, discovered that they had become the objects and not the subjects of the Plan. In taking negativity as method, Cacciari and Tafuri try to learn from that experience, seeking a means to equal and outrun the effects of the Metropolis and Plan. Imagining that one's critique could occupy a position external to the Metropolis—and remain untouched by its destructive logic—is tantamount to being oblivious to some of its most powerful effects.

Cacciari believed it was necessary for critical thought to immerse itself in the powerful currents with the aim, somehow, of riding them. We require a self-conscious approach that understands how to "integrate the negative within itself."[113] As he puts it:

> The negative stays within the limits of the Metropolis, since it has uncovered the Metropolis's negativity. But this negativity, once demythified, demystified, and thrust whole into Erlebnis and Verstand, presents an image of the Metropolis as symbol-place of the contradictions and functions of modern capitalist society. The negative, used correctly—that is, according to the terms of its own hopelessness, and not mystified as a requisite for synthesis, as a prayer for consolation—leads to this limit.[114]

The seizure and embrace of capitalist negativity, advocated here, is meant to reveal its contradiction and its limits. This is the strategy of completed nihilism, which lives through the processes of disenchantment to a point at which it is, arguably, beyond nihilism, or, at least, in excess of it. If, according to Nietzsche's well-known definition, nihilism is the devaluing of the highest values, then completed nihilism is the point reached when this logic of devaluation is pursued through to fulfillment.[115] Completed nihilism, as Nietzsche puts it in *The Will to Power*, requires a strength of character that will "deify becoming and the apparent world as the only world."[116] Only on the basis of this presupposition and acceptance of total disenchantment—only by having the strength to abandon all traditional and transcendent values—is it possible to engage actively in the creation of values appropriate to the current period. In Nietzsche's terms (and terms which take us back to Jameson's and Clark's charges):

> It is only in this sense that we are pessimists; i.e., in our determination to admit this revaluation to ourselves without any reservation, and to stop telling ourselves tales—lies—the old way.
>
> That is precisely how we find the pathos that impels us to seek new values. In sum: the world might be far more valuable than we used to believe; we must see through the naiveté of our ideals. [117]

This revaluation would consist of values that emerged immanently from the current conditions of modernity.

It is a matter of some debate whether Nietzsche conceived completed nihilism as, strictly, the final, all-embracing phase within nihilism or as one beyond it. In *The End of Modernity*, Gianni Vattimo adopts Heidegger's view and challenges the idea that completed nihilism represents the "overcoming" of nihilism.[118] On his reading, nihilism's very demolition of foundations makes impossible all acts of "surpassing" and predicates against the very possibility of new values being created. The postmodern condition for Vattimo is coextensive with this realization of foundationlessness, which is also completed nihilism. Only the wish to be beyond nihilism can be overcome by adopting a different attitude to the situation.

Tafuri and Cacciari shared the critique of overcoming (and, as we have seen, the emphasis on attitude); it was necessary, Tafuri argued, to take Nietzsche and Heidegger seriously.[119] But they also put in question the metaclaims—themselves inspired by forms of neo-Nietzschean philosophy—to have exceeded modernism. Postmodernism—and Tafuri had in mind its high expressions (semiotics, rhizomatics, and Derridean difference) as much as the popularized accounts that were particularly visible and stylized in architectural history—tended to obstruct questioning by presenting matters as resolved, as if we were "already *beyond (post)*," naively imagining that it had surpassed contradictions and thus falling prey to a type of complacent substitutionalism.[120] According to Cacciari "nihilism cannot do without solution"; the idea of going beyond would be a necessary accompaniment to devaluation, and this aporia forced nihilism "to mythify its own immanent will to power."[121] All the "hypermodern 'dances,'" Tafuri noted (of the sort of postmodernism promoted by Paolo Portoghesi), were just nihilism in its "incomplete" or "imperfect" form.[122] With its relapse into the illusion of making easy syntheses of form and meaning and its banalization of philosophical themes into Disneyland logic, postmodernism obstructed "not only radical questioning, but all questioning." The task instead was "how to question *radically*—without entertaining any useless illusions about recomposing what has been shattered or resynthesize the plurality—an era that no longer permits an agreement between wholeness and multiplicity, but also does not oblige one to comply with its most recent victors."[123] Portoghesi, of course, was not Vattimo—as Tafuri well recognized—yet Tafuri's closing comment criticized even postmodernism's more articulate champions. Before the postmodern ascendency, Tafuri and Cacciari already considered questions beyond what were to become postmodern philosophy's own self-inscribed circle of concerns. Thus, in Cacciari's words, the fulfillment of completed nihilism would mean "neither the task of effecting solution nor that of effecting the end of all solution."[124]

Despite the overt hostility of Tafuri and Cacciari to the Aufhebung, we can see that their argument nevertheless echoes Hegel's claim that thought truly finds itself through the process of its "utter dismemberment," through the fall into the depths of its own hopelessness.[125] For all the antagonism toward the "Hegelian synthesis," their argument, nevertheless, accepts many of the dialectic's tropic turns and transitory characteristics, its movements of internalization, integration, introjection, and immersion. Their version of completed nihilism requires the strength to face capitalist negativity: this is, in Calvino's terms, acceptance of, and immersion in, the inferno. However, for Cacciari and Tafuri, this immersion in the currents of the Metropolis is the necessary condition for seeing the Metropolis's shape and the only means to avoid being blinded by its effects. Going into, and through, nihilism would provide a strategy to enable us, *pace* Calvino, to see "who and what . . . are not inferno" so that those forces could acquire "space." The surrender being advocated, then, is of a specific sort. As Tafuri notes of Oskar Schlemmer, his act of affirming the Metropolis—the "great yes" said to reality— had a double effect, both confirming the reification of human life and yet, precisely through this affirmation, acting in the cause of liberation.[126]

Returning to a passage that we have already encountered, the crucial shift can be clearly recognized: "*To save oneself* one must lose one's self, one must resign oneself to being submerged in the chaos, one must make oneself sign among signs. But *by action*."[127] The affirmation of nihilism—acknowledging this world as the only world and calling it "good," as Nietzsche put it—forms the prelude to a salvaging of an emacipatory mode of negativity, one that goes beyond negation's appropriation by capital. (Indeed, it is notable that Cacciari and Tafuri do not follow Deleuze who, in his return to Nietzsche, rejects negation and dialectic *in toto*.)[128] Only by working with reification's "geometric freezing of the real" would it be possible to find a "space of action" and "liberty."[129] Negative thought, according to Cacciari, is the force that "lays bare the logic of this society . . . the negative reaches the point where it exposes this society's internal conflicts and contradictions, its fundamental problematics or negativity."[130] In addition to claiming a space of freedom and action, negative thought, it is argued, could break through the "silence" of the *nihil*, the complete internalization of Metropolitan nihilism allowing the "tragedy of the given" to "speak for itself."[131] Tafuri echoes the same point in the context of a consideration of historical method, although he takes a more forceful approach to this moment of ventriloquization, suggesting that negativity must be *made* to speak: speak of its making, its becoming, its function, and its conflicts:

> Historical space does not establish improbable links between diverse languages, between techniques that are distant from each other. Rather it explores what such distance expresses: it probes what appears to be a *void*, trying to make the absence that seems to dwell in that void speak.[132]

The space spoken of by Calvino here becomes a presence in historiography. What, in the strategy of completed nihilism, appears at first sight to be a collapse of critical perspective is thus turned around. It is by making silences into "determinate abstractions" that the historian establishes "historic space," Tafuri notes. "Distance is fundamental to history"; indeed, the historian "must create *artificial* distance"; or, as he remarked of the relation between architectural theory and practice, "it is the conflict of things that is important, that is productive."[133]

Contro il piano

Writing in 1975, in his preface to the English edition of *Architecture and Utopia*, Tafuri observed that his 1969 essay "Towards a critique of architectural ideology"—the early version of the book—had been subject to much criticism (to "many more or less violent stands") and, more to the point, he argued that it had been grossly decontextualized:

> The journal that published this essay (and others by myself and by colleagues working along the same lines) was so clearly defined in its political history and particular line of thought and interests, that one would have supposed that many equivocal interpretations might *a priori* have been avoided.
>
> This was not the case. By isolating the architectural problems treated from the theoretical context of the journal, the way was found to consider my essay an apocalyptic prophecy, "the expression of renunciation," the ultimate pronouncement of the "death of architecture."[134]

This theoretical context—the journal in question—was *Contropiano: materiali marxisti*. In addition to this early essay version of *Architecture and Utopia*, the journal was the forum for Tafuri's discussion of the Austromarxist urban projects of "Red Vienna" as well as Cacciari's early articulations of negative thought.[135] Most of Cacciari's essays in *Contropiano*, however, had more explicitly political subjects: "Capitalist development and the cycle of struggle"—a consideration of industrial relations at the Montecatini-Edison plant in Porto Marghera—while his essay on political theory and organization in post-1968 France, represented the first sustained analysis of the May events on the Italian Left. Launched in 1968, *Contropiano*'s founding editors were Cacciari, Antonio Negri, and Alberto Asor Rosa, a lineup indicating that the journal indeed provided a very specific context and was directed toward Marxist intellectuals of a distinctly workerist persuasion. Its editors and contributors, such as Mario Tronti, were among the most prominent theorists of *operaismo*.[136] In the Italian context, "workerism" did not have the same dismissive ring as it does (or did) within places such as the United Kingdom, where it implied forms of Left-wing politics that followed, or even aped, patterns of immediate consciousness among workers rather than attempting to challenge or transform them. Italian *operaismo* emerged in the nineteen-fifties as a distinct anti-Stalinist tendency among dissidents of both the PCI and the PSI. As a student at the University of Padua, Cacciari had encountered the group with whom he would later edit *Contropiano* in the milieu of *Quarderni Rossi* and *Classe Operaia*, respectively the formative and fully fledged publications of workerist political analysis. Cacciari credits Negri—who consolidated his teaching position at the University of Padua in 1967 when he became Chair of State Doctrine—with encouraging him to read classic philosophical texts by Kant and Hegel, as well as introducing him to Tronti and Asor Rosa, the key workerist figures in Rome.[137]

Workerist intellectuals tried to rethink Marx's theory of political economy in the context of their experience of Italy's postwar "economic miracle." The extraordinary pace of modernization and urbanization through these years saw large-scale migrations and the continued restructuring of labor relations. The workerists also put a strong emphasis on bringing together questions of theory and political practice. Like the unorthodox Trotskyist tendencies "Socialisme ou barberie" in France or the *Correspondence* group of Raya Dunayevskay and C. L. R. James in the United States, and like the Situationist International too, *operaismo* drew its political lineages from a critical engagement with council communism and the experiences of early Weimar revolutionary activism, the phenomena that Lenin had famously criticized as "Left-wing communism."[138] Like these other tendencies, *operaismo* turned its attention to the problem of the labor movement's bureaucratization and to the autonomous power of the working class. (This was one of the political contexts from which Italian Autonomism emerged in the nineteen-seventies, and remains at the core of many of the ideas that, more recently, have been rehearsed in the books *Empire* and *Multitude* by Antonio Negri and Michael Hardt.) Developed by Panzieri and Mario Tronti, workerist theory argued that the ideas of the established Left—the PCI and the PSI—had stagnated into static, objectivized, and economistic categories, which inflected how the Left operated within the capitalist state (for example, the means of trade union struggle adopted) and determined its idealization of the productivist ideals of Stalinist socialism. In their view, the mainstream Left too readily accepted the framework laid out within capitalism, failing to challenge categories such as work and production; accordingly, they argued, this old Left treated the working class as merely a defensive, "reactive" element within the labor-capital relation. As central players in the New Left's "return to Marx" and in the recovery of the labor process as a site of theoretical and political activity, the Italian workerists placed the working class as the active component.[139]

These thinkers were also keen to understand the changes in class composition, appropriating sociological methods for their own "militant research." In particular, they attended to the emergence of what they called the "mass worker," the unskilled or semiskilled employee who—through the restructuring of capitalist production, and the progressive impact of Fordist and Taylorist methods—had displaced the centrality of skilled operatives in the production process. For the official trade unions, this downgrading of skill was associated with the diminishing of their bargaining power. Their response was to focus their efforts on a defense of existing grade hierarchies. Where the traditional labor movement saw these developments as a threat, workerists believed that this leveling of the proletariat (and the qualitative aspects of their work) had a positive side; it led them to identify fresh political potential in the increasing homogenization (or the dehierarchizing) of the new "social mass." (Indeed, "massification" not only was understood as the equalization of skill and grading among a traditional workforce but also embraced the new layers of female and migrant labor; male prejudice prevailed, but workerism was not blind to difference.) As Tronti argued, massification was not simply the quantitative accumulation of exploited workers, but a process of growth and internal homogenization of industrial labor power.[140] Massification, it was thought, clarified the struggle against the capitalist organization of work per se; collective bargaining was the process through which massification of the class could translate into massification of struggle. This mass worker—who existed in workerist discourse somewhere between an immediate reality and an ideal toward which reality was tending—increasingly lacked any ideological investment in, or attachment to, their work. In a way that was clearly paralleled in Tafuri and Cacciari's argument about Metropolitan reification, this mass worker was understood to be becoming totally alienated.

> If this corrupted word [alienation] still has a meaning, it is only that of expressing a specifically determined form of direct exploitation of labor on the part of capital: *total* estrangement of labor with respect to the worker; useful, concrete labor which becomes objectively estranged, external and indifferent to the worker; the end of the trade, of the *profession*, of this last appearance of *individual* independence of the worker, the extreme survival of a bourgeois *person* in the body of the worker.[141]

The consequence, the proponents of *operaismo* thought, would be progressively consolidated in a "massified" contradiction to capital. The central issue for them was how to transform this growing structural autonomy into a force of political negation. In order to focus the potential whereby "living labor" might reject its own commodity status, the strategy of *operaisti* concentrated on disrupting the operations of "labor power," that is, labor in its commodity form. They promoted tactics like the go-slow, industrial sabotage, and organized resistance to productivity deals; they also developed arguments for a social wage. One of the key reasons for their growing influence was their success in helping worker-militants to disentangle their desire for wage increases from managerial efforts to raise productivity.

In various regions workerism was dominated by different groupings and tendencies: Potere Operaio in the Veneto, Lotta Continua in Turin, Il Manifesto in Rome, or Avanguardia Operaia in Milan. By the late sixties, workerist ideas had acquired a large base of support among intellectuals and industrial workers, with particularly strong bases in the Turin car works and in the petrochemical plants that lie just across the water from Venice in Porto Marghera. Cacciari, along with Negri and Paola Meo, had organized *Capital* reading groups in the petrochemical plants of the Veneto in August 1963; and, the following year, they had tried to establish rank-and-file committees that would be independent of the official unions. As Paul Ginsborg has observed, "the Italian protest movement was the most profound and long-lasting in Europe" and the "Italian revolutionary groups, taken together, were the largest new left in Europe."[142] In the period leading up to the events of 1968, the Veneto caucus was one of the most important of these groups, and by late sixties and early seventies it dominated the Venetian far Left and became a primary focus for critical Marxist intellectuals. In 1967, a series of research seminars, initiated to commemorate the fiftieth anniversary of the Russian Revolution, took place at the Institute of Social and Political Science at the University of Padua. At the end of the year, a conference on international class struggle was organized, a forum from which emerged some of the canonical texts of Italian workerism, *Operai e Stato*.

Constant tensions within *operaismo* led to a number of organizational splits. At the time of the launch of *Contropiano* in 1968, the editorial board consisted of individuals who operated both inside and outside the PCI. The fractures among this group of intellectuals turned on two moments. The first of these was when the Rome caucus around *Classe Operaia* (Tronti, Asor Rosa, and Rita di Leo) decided to enter, or reenter, the PCI in 1967. This moment also resulted in the founding of Potere Operaio veneto-emiliano (POv-e) by Negri and Cacciari, the Veneto-based opponents of this strategy. Subsequently, Negri and Cacciari parted ways, with Cacciari taking Tronti's path and Negri leaving the editorial group of *Contropiano*, a situation that came to a head over a dispute over Tronti's article for the journal's first issue.[143] The tensions over organizational independence were to recur throughout the history of *operaismo*, and, like so many political fractures, produced bitter antagonisms. Nevertheless, it would be incorrect to deduce that those who entered (or never left) the PCI abandoned all the core themes of *operaismo*. The attitude of *operaisti* to the PCI swung, as Steve Wright has noted, between Franco Piperno's dismisal of "the working-class articulation of capitalist social organization," as he characterized the PCI in 1969, and Asor Rosa's decision to exploit PCI forums as "transitory meeting places for revolutionary militants."[144] The primary tension in workerist strategy was between, on the one hand, its commitment to political autonomy and, on the other, the need to build an organization that was capable of advancing struggles without unnecessary exposure to defeat. Yet, even well into 1968, the relation of the Veneto workerists to the use of the PCI in general, and to Tronti's decision in particular, remained ambiguous.[145] As a founding theorist of *operaismo*, Tronti was still influential on those outside the PCI, as to a certain extent was Negri on those inside. While the POv-e group considered the PCI in Emilia-Romagna a lost cause, they saw the Veneto communists as more open to influence, the 1968 Venice conference "Students and Workers" being co-organized by POv-e and the PCI. (It was in the wake of this event that Cacciari abandoned POv-e for the PCI.) However, as the struggles escalated in the factories later the same year, workerists and PCI activists clashed on picket lines, which further polarized opinion, and confirmed for many militants that even the Veneto PCI was beyond reclamation. The group associated with *Contropiano* survived this period of difficulty, as too, it seems, they did the PCI's efforts, in 1969, to purge itself of Left-workerists. These internal divisions provide the backdrop for the *Contropiano* project and for Tafuri's arrival in Venice. When Tafuri alludes to the importance of the journal for understanding his project, there is rather more at stake than a simple intellectual context. Tafuri was a partisan of Tronti's wing. For his 1970 symposium on the migration of European architects to the U.S.S.R. in the late twenties, Asor Rosa and Rita di Leo were among the invited speakers.[146] Speaking on topics relevant to the theme in its most broad sense, their role was designed to strengthen the theoretical punch of the occasion, but, more especially, to secure its political direction. Although Cacciari departed from POv-e, it would be foolish to conclude that the ideas associated with *operaismo* disappeared from his work. Although he attempted to theorize and legitimize his change, Cacciari joined the PCI while retaining important aspects of the workerist project to understand class composition, and was opposed to Leninist models of political organization. Similarly, Tafuri's research interests and theoretical arguments, as we will see, continued to explore workerist themes. Many in the PCI were workerists in spirit, existing inside an organization that they fundamentally distrusted and within which they were often equally mistrusted.

Formal histories of workerism suggest that it remained focused on blue-collar workers and was little concerned with student activism until the middle of 1968, and that the interest of the student movement in class politics and of workerists in students dates from the moment when the Movimento Studentesco made its "turn to the class." However, discontent in schools of architecture, along with faculties of medicine, was a feature of Italian higher education from the early sixties. Negri has recalled how IUAV "had been a center of student resistance since 1965. . . . In 1968 students from Venice and Padua joined forces with the workers at Porto Marghera. This worked out quite smoothly because they had been in constant contact for a decade: the school of architecture was a gathering place for the working class."[147] IUAV had been occupied in 1958 and students were closely involved in antifascist demonstrations in 1960. Its students and librarians had been involved in the defense of the workers at Sirma in Marghera in Spring 1965, a struggle that is considered as the initiation of cooperation between workers and students.[148] Another occupation in 1967 adopted the general assembly, a form of direct self-government that rejected representational structures, which became the model for industrial militants. Particular issues were generalized; in March 1968 the students argued that there would be no harmony in the universities without harmony at Montedison and that theirs was "la lotta contro il piano generale dei padroni" ("the struggle against the bosses' overall plan"). In June, when the convention "Students and Workers" was held in Venice, groups of worker and student militants blockaded the opening of the Biennale (it opened three months late at another venue); they later disrupted the Mostra. (A couple of years later, in a way that was to be bizarrely echoed by Cacciari's part in the 1999 counterpublicity campaign, these groups blocked the routes of industrial flow, including the transport routes of German tourists.) The intellectual collaborations for which the school of architecture became known were forged in the intense atmosphere of debates—theoretical and practical—at the conjuncture of intellectual enquiry and revolutionary militancy. Debates on radical town planning and the housing question also had urgency, not only in Venice; schools of architecture had been occupied in 1963 in opposition to government policy, and there continued to be widespread challenges to the assumptions and hierarchies of professional training and university pedagogy. All this came to a head in 1967: engineering students challenged the ideas of technical rationality and efficiency at the heart of their profession; journalists argued against the notion of impartiality, young lawyers against the supposed neutrality of the courts. Graffiti in Milan criticized an architectural professor: "Bonicalzi, you who love prefabrication, tell us about building speculation."[149] It was unusual for members of academic staff to sympathize with the new mood, but, in Venice, the political critique of professionalist ideology in architecture chimed with the students' concerns.

None of the projects initiated by Tafuri, or the lines of argument developed through them, can be adequately comprehended apart from the broad political project shaped by the intellectual culture of *operaismo*, and his assessment of development leaves little doubt as to his leanings. This orientation is also apparent in his position on industrialization in the U.S.S.R. (specifically his criticisms of the Soviet emphasis on industrial production and its cult of work) and his adoption of the argument on state capitalism. Tafuri and Cacciari's interest in work on Yevgeny Preobrazhensky, the economist of the Left Opposition, provides further evidence, Preobrazhensky's account of "permanent revolution" offering an early and radical articulation of an economy of dynamic process or, to use Tafuri's words, "a theory of the plan based explicitly on dynamic development, on organized disequilibrium, on interventions that presuppose a continual revolution of mass production."[150] (In 1971, a special issue of *Contropiano*—including essays by Cacciari and Dal Có—was devoted to Soviet industrialization.) Tafuri's approach to the avant-garde "machine-man," even in *The Sphere and the Labyrinth*, develops themes from Panzieri's famous 1961 essay on the "The Capitalist Use of Machinery."[151] The project on the American city, with its emphasis on the New Deal, is another direct outcome of workerist research, seeming to offer an example of a society that was free from nostalgic illusions for earlier forms of capitalism. Considered to be of crucial importance within this strand of politics—as still in Hardt and Negri's postworkerist and postautonomist *Empire*—the United States demonstrated the processes of capitalist modernization and the emergence of social capital, in the absence of an effective social-democratic tradition. In his 1971 postscript to the second edition of *Operai e Capitale*, Tronti summarized the position: "the American class-struggles are more serious than European ones in that they obtain more results with less ideology."[152] The American working-class struggles of the nineteen-thirties—"this red sun that comes from the West"—represented the way forward for the "new politics," he continued.[153] The political issue at the epicenter of working-class history was not simply reform versus revolution—the frame within which Tafuri's arguments have so far been set—but the different attitudes represented by the European and American struggles. The barrier of ideology—central to Tafuri's argument—included prominent traditions of Marxism too; the orthodox Marxism of the PCI and PSI had been inadequate, mediating class struggles in a manner considered typical in the European experience. The opposite was seen to be the case in the United States, Tronti thought; while Marxism was far weaker there, and only experienced "indirectly," the American situation had nevertheless proved—and this, of course, was a highly controversial thesis on the Left—to be more "objectively Marxian."[154]

However, it is the notion of the "Plan" that best situates Tafuri's argument within the workerist political frame. The Plan, was itself premised on the socialization of capital, and what Tronti saw as the concomitant socialization of class struggle "which arises from immediate needs of production and reproduction of social capital"[155] Even within the context of *operaismo*, Plan was particularly prominent to the language of POv-e. The first issue of POv-e's paper in 1967 included an article "Contro il piano," one of a series of similar headlines.[156] The factory, POv-e argued, was the site where capital's plan was most clearly manifested, "and from whence it draws its strength." The worker-student unity of 1968 launched, as it was put, "an open and general struggle against the entire plan of capital."[157] The title of the theoretical journal condensed this stance to *Contropiano*; "Against the plan" became the snappier "Anti-Plan." The contraction was both more abstract ("plan" lost the referential specificity of the definite article, even if that specificity was as loose as "il piano dei padroni") and more ambiguous (*Contropiano*'s editors were against capitalism's "Plan," but now understood their political task to be the construction of an alternative or counterplan).[158] In his essay on "The Capitalist Use of Machinery," Panzieri had outlined how capitalist planning expanded beyond the locus of the factory, where capital tried to integrate the different aspects of the labor process: "the progressive extension of planning from the factory to the market, to the external social sphere."[159] Thus, the concept of the plan was quite broad in is scope, and Panzieri's description from 1961 already anticipates the later workerist idea of the "social factory." Revolutionary struggle, Panzieri argued, was the "global opposition to the capitalist plan"; the point was not simply to "acknowledge and exalt" technological rationality, but to develop a "dialectical awareness" of both the "technical" and "despotic" moments in the capitalist organization of production, and to subject technology to "the socialist use of machines."[160] At stake in all these discussions of the plan, then, is not only the prospect of Italian modernization, but also the very concept of modernity.

The planning of economic development had been at the heart of postwar political alliances and, in 1962, formed the platform for the Center-Left government coalition between Christian-democrats and the PSI. The associated planning and building programs, according to Robert Lumley, "were touchstones" for the Centre-Left experiment, their progress being closely monitored by architectural students.[161]

Christian-democratic policy was essentially a form of free-enterprise capitalism; so, it was the socialists who invested their hopes in reform, believing that their coalition would facilitate substantial improvements in the life conditions of their voters. The Left hoped to go beyond what they called mere "corrective" reforms and to achieve "structural" changes. The idea that planning provided the route to socialism was put forward in the writings of the PSI's Riccardo Lombardi and was a view largely shared by Togliatti's PCI. It became increasingly clear, however, that the much-touted improvements were either failing to materialize, or, alternatively, that the plans were being conducted in the interests of capital. The unions themselves had put forward the Piano del Lavoro in 1949, but even the moderate nature of its neo-Keynesianism was not enough to stop it being quashed. By 1964 there was extensive disillusion that reform could provide improvements, let alone travel an "Italian road to socialism." In Ginsborg's assessment:

> there was very little evidence to show that such reforms could be achieved in Italy without engendering a major confrontation with the ruling élites. The experience of the centre-left had, on the contrary, given ample warning of the capacity and vigour of the conservative elements of the Italian ruling class. They had successfully sabotaged reforms far more moderate than those Togliatti had in mind.[162]

This realization spurred the growth of dissident workerist groups. For many on the Left, the PCI—which, by way of "constructive opposition," had tried to make itself into another potential partner in the governmental coalition—did not appear to offer a focus for resistance. In the postwar decades, the PCI had agreed with the ruling governments that the modernization drive was vital to retaining international competitiveness. Moreover, for the PCI, economic and technical development was seen as distinct from the condition of social relations. Thus, to draw on Marx's terms, they saw the social relations of production as shackling an otherwise positive growth in the forces of production. Techniques, machinery, and issues of productivity were seen as neutral factors lying outside the political arena. Panzieri, in contrast, had argued that technique was inseparable from the social relations of production and the capitalist organization of labor, arguing against the objectivists who, as he put it, subscribed to "a technologico-idyllic image."[163] In other words, supporters of *operaismo* emphasized the politicized nature of technique, production, and development; economic planning was also deemed to be inseparable from the politics of restructuring and the socialization of labor. Thus, the English subtitle to *Architecture and Utopia*—"Design and Capitalist Development"—does not merely signal some general Marxist-inspired interest in locating art and culture within the context of the capitalist economy; it carries over specific resonances from Italy's far-Left political culture.

Occupying a particularly important role in the politics of reform, Tafuri takes the question of mass housing to be an exemplary form for modern architectural experimentation.[164] Accommodation was an issue that was felt with some immediacy by a large part of the population in Italy; there was widespread poverty and homelessness, which was exacerbated by the waves of inter-regional and internal rural-urban migrations. The problems of urban reform were evident in the films of Italian neorealism, from de Sica's *Miracle in Milan* (1950) and Fellini's *The Nights of Cabiria* (1957) to Pasolini's *Accattone* (1961); they took a more indirect—and a less picaresque, less picturesque, and less sentimental—form in Tafuri's writing, where he addressed the large-scale housing projects developed under social-democratic governments or administrations, in places such as Weimar Germany, Amsterdam, and Red Vienna.[165] Indeed, it was the limitations and defeats of these projects that concerned Tafuri, in which he found echoes of the experiences and disappointments of postwar Italian architects, such as Ludovico Quaroni and Mario Ridolfi, who, in the fifties, had played a major role in the design of workers' homes. (Quaroni had been Tafuri's teacher as well as the subject of his first book.) It seems that his interest was also, in part, a reckoning with the PCI's pride and joy: their control of Bologna's city administration. Especially in the nineteen-sixties, the PCI emilia-romagna had done more than any other city administration to implement the government-initiated *Piani di Edilizia Economica e Popolare* ("Plans for Economic and Popular Housing"). Like Ernst May in Frankfurt during the twenties, the Bolognese council had some significant successes. In Tafuri's view, the crisis that had emerged for the Frankfurt project was not identified with the rise of the political Right, so much as with the economic sea in which the experiment floated, and which, as he put it, "came to grief" on the rocks of the 1929 crash. From this perspective, Bologna's social-democratic reform was little different; political control by the PCI did not equate to economic power, and many of their reform plans were thwarted by this disjunction. Unlike the Frankfurt experience, the PCI held back from measures to socialize land, anticipating the likely reprisals. Its attempts to create neighborhoods and community—in an effort to stave off social atomization and alienation—also echoed with the interwar projects of the Viennese *Höfe*. Moreover, the attempt to replicate the social relations of the village in the city—whereby organic relations would be reproposed as isolated complexes "fortressed" from the Metropolis—was taken by Tafuri to be not only retrograde but also incapable of success within its own terms. The Metropolis, as we have seen, would allow no such protective zone and would simply consume these concrete "prayers for consolation." Taking a stronger line than Tronti, Tafuri's argument while clearly unpalatable to many on the Left—then and now—was certainly no dreadful "nothing is possible" approach; it was a hard reckoning with the contemporary politics of the PCI.[166] The reformist policies of the PCI's "showpiece," however indirectly, represented an important test case for a workerist historian and theorist of the Metropolis and the Plan.

Tafuri's position was clearly critical of the PCI and PSI (although he was a member of the former), but it is important to recognize that it also represented an intellectual challenge within the New Left.[167] Much as his politics would have appealed to the student mood at IUAV and beyond, his arguments stood in sharp contrast to the widespread fascination with Maoist cultural revolution and with populist desires to serve the people, which were especially prominent in the fields of theatre, medicine, and architecture. Lumley describes this populist turning in terms that, in the context of our exploration of Tafuri's theory, are remarkable poignant:

> The disorder of the city was rejected in the name of an ideal, organic society which was rural in its inspiration. It was utopian, therefore, in the pejorative sense, because it attempted to deny the complexity, diversity and conflict of modern societies.[168]

Readers of Tafuri and Cacciari—and their account of the Metropolis—will recognize the stakes here. Tafuri was certainly not tailcoating some broad Zeitgeist among the younger generation; rather, he was consciously intervening, from a specific radical perspective, in contemporary Left-wing political debates. In criticizing the approaches of both the Old and New Left, his position was unafraid of confronting difficulties, whether those of political difference or the type of challenges demanded by focusing on the problem of form.

An influential source for Tafuri's arguments concerning the Plan in *Architecture and Utopia* was an essay by Negri entitled "Keynes & the Capitalist Theory of the State post-1929," which was published in the first issue of *Contropiano* in 1968.[169] In this article on John Maynard Keynes, Negri attempts to elaborate the historical transitions outlined by Tronti in *Operai e Capitale*. Negri periodizes modern state development according to the waves of class struggle and revolutionary upheaval: 1848–1871 representing the period of rising working-class autonomy; 1871–1917, the phase when that autonomy gained organizational form; in particular, he concentrates on the interwar period, which, he argues, saw the rise of a new state form that had been prompted by the events of 1917. The Russian Revolution, he suggests, did not just pose an external threat that could be dealt with through military intervention and diplomatic isolation. With the explosion of new mass trade unions and factory councils across Europe, Negri argues, the Revolution heightened the internal threat to capital. In response, capital transformed its own substantive content, and repression took a "technological path" as a means "to undermine the material base" of the working-class vanguards. Hence, as the productive process was massified and the labor force deskilled under the impact of Taylor and Ford, capital restructured itself, forcing a "leap in organic composition in new sectors; assembly line; flow production; scientific organization of work; sub-division/fragmentation of jobs, etc."[170] Picking up on Tronti's emphasis on the New Deal, Negri suggests that capital's response to 1917 only became clear after 1929. For Tronti, this period represents an important phase in the class struggle, one in which workers responded, in turn, to capital's own structural revolution, its own adoption of continuous revolution.[171] This was the moment, according to Negri, the dominant tendencies in the ruling class finally abandoned nostalgia for the liberal constitutional state, facing up to the loss of values—that is, the overthrow of values by the protagonists of 1917.[172] Keynes was identified as the economist who most thoroughly accepted the implications of this disenchantment and who provided the most advanced economic science for the new phase within capitalist social relations.

For Negri—as for Tronti and Tafuri—the significance of Keynesianism at this point should be understood not as the initiation of state intervention policies, but rather as "the reconstruction of a state based on the discovery of the inherent antagonism of the working class." In Negri's view, "new weapons are forged in order to prevent the [working] class acting outside capital, and to make it act within a framework whose outlines are continually being drawn anew."[173] In the wake of 1929, 1917 finally "manifested itself in a crisis of the entire system."[174] The threat posed by the working class had to be sublimated by capital "into a dynamic element of the system" and absorbed "at ever new levels."[175] The state became simultaneously "planner state" and "crisis state," operating within civil society and continuously recreating "the source of its legitimacy in a process of permanent readjustment of the conditions of equilibrium."[176] Unlike the "socialist reformism" of the Left, this "capitalist reformism"—and here we hear strains of Nietzsche—did not "whine" about the imbalances of the system; rather it asserted its own class interest, and sought to resolve the difficulties through its own self-reproduction. Tronti also distinguishes the "reformism of the labor movement."[177] Reflecting on the period of struggles opened by the New Deal, he notes that:

> The great capitalist initiative has been a labor victory if only in that it allows us a crude knowledge of the enemy at its highest historical point. Afterwards it becomes useless to condemn it: our only advantage is in using it.[178]

Here Tronti and Negri articulate a position with clear echoes in Tafuri and Cacciari: from the critique of ideology (including that of official Marxism), to the rejection of attitudes of tragic anguish and the advocating of completed nihilism.[179] As Negri famously puts it, capital had learned from *Das Kapital*; "for its own self-preservation" capital had claimed "permanent revolution" but by way of "a paradoxical *Aufhebung*."[180] In this formulation we can recognize the parallels between the workerist account of capital's internalization of proletarian negation and Tafuri's account of the incorporation of avant-garde; we now turn to consider how this has been misunderstood.

Without Mediation

> The dialectic is finished. Hegel is dead. What remains of Hegel is the self-consciousness of the bourgeois world. The bourgeois world is dialectical and cannot but be dialectical. But we are not. The workerist critique is not today the restoration of the dialectic, but rather the discovery of the terrain and the form of the conflict.
>
> —Toni Negri[181]

> We do without mediation because of our hatred of opportunism.
>
> —Mario Tronti[182]

The critique of the avant-garde, the Left-Nietzscheanism and embrace of completed nihilism, the opposition to Universal History, and the rejection of dialectical synthesis: these factors might sound like yet another account of the route away from the radicalism of the sixties. Most accounts of Tafuri, however, fail to recognize the unique intellectual context of *Contropiano*; in so far as they detect its echoes, they hear just one part of the dynamic outlined in the workerist theory, the moment where negation's radical impulse is overtaken by capital (when capital learns to read *Kapital*, to recall Negri's formulation). To arrive at such a conclusion, these readers have to ignore a central precept of workerism: its assertion of the primacy of living labor over capital, that is, its emphasis on workers' role as the active and the determinate force in the struggle with capital. This argument reverses the longstanding emphasis on the centrality of capitalist development, to which the class reacts. According to Tronti: "At the level of a fully socialized capital, capitalist development becomes subordinated to working-class struggles, it follows behind them."[183] Negri merely echoes Tronti's famous "inversion formula" when he states that the working class became "the decisive element and the motive power" for development.[184] Furthermore, the "ineliminable antagonism" of the working class to capital is seen not only as emerging in the big showdowns of mass industrial action or in the dramatic highpoints of struggle (the stuff that, for many histories, is what it takes to signal the very existence of social resistance, and what makes possible the entrance of categories such as class into representation). The workerist perspective considers resistance to be both a permanent and universal feature of capitalism, operating in the daily refusal to submit to workplace discipline; such small and ongoing refusals taking place even during periods of apparent quiescence.[185] As Tronti puts it:

> Capital's power appears to be stable and solid . . . the balance of forces appears to be weighted against the workers . . . and yet precisely at the points where capital's power appears most dominant, we see how deeply it is penetrated by this menace, this threat of the working class.[186]

In his essay, Negri takes the very existence of the working class to be "the sign of a latent dissociation," for he considers it to be not just "inside" capital but "also capable of stepping outside it."[187] Thus, in contrast to other forms of Western Marxism, an acknowledgement of capital's internalization of negation did not presuppose the collapse of that resistance, let alone conclude in arguments about "the end of the working class." The most that capital can expect from its absorption of negation and its development of new modes of repression was a postponement of its problems. Indeed, in workerist theory, as Negri's account of Keynes makes clear, it is living labor that endows capital with—or obliges it to acquire—its modern dynamic forms, its "permanent revolution." Twentieth-century history, from a workerist perspective, is not to be understood as capital's expanding imperium over resistance, critical distance or negation; or rather, it is not to be understood solely as that process: the full dialectic had to be grasped as the political process of social struggles in which working-class latent dissociation was primary. Accordingly, in *Architecture and Utopia* Tafuri argues that merely recognizing the "negative is inherent in the system" is insufficient, the critical task being to discover what "actually makes this 'negativity' (the negative of the working class) function as a 'necessity.' "[188]

The political project of workerism was to ensure that one side of the dialectic, the *via negativa*, broke "free into independent destructive action."[189] Capital's appropriation of permanent revolution was to be met with "a permanent continuity of struggle," which would take "ever-new forms" devised by the "intellectual creativity of productive work."[190] Still deploying Hegelian terminology in 1964, Negri poses the problem in terms of the contrast between two types of Aufhebung: the capitalist Aufhebung holds down the working class but cannot be a true sublation because it fails to "overthrow its object"; in contrast, the working-class Aufhebung "burns the successive levels of capital's development, reproposing its rupture and its supersession."[191] Although generally more cautious in his formulations than Negri, Tronti's writing is nevertheless replete with the language of rupture: workerism needed to combine the "patience of research" with "the urgency of the answer"; it needed to turn everyday class recalcitrance into "the Strategy of Refusal." "Without mediation," Tronti writes, the struggle could develop as "a rapid clash between immediately contradictory concepts" that will force the movement "to jump ahead."[192]

> Today, at the theoretical level, the workers' viewpoint must be unrestricted, it must not limit itself, it must leap forward by transcending and negating all the empirical evidence which the intellectual cowardice of the petty-bourgeois is forever demanding. For working class thought, the moment of discovery has returned. The days of systems-building, of repetition, and vulgarity elevated to the status of systematic discourse are definitely over. What is needed now is to start again, with a rigorously one-sided class logic—courage and determination for ourselves, and detached irony towards the rest.[193]

As Cacciari echoes, writing a few year's later in "The Dialectics of the Negative and the Metropolis," "our progression should be from the negative, to the Metropolis as an instrument of class, to its negativity as a contradiction of class: from the perspective of the negative to the perspective of class."[194] In this sense, then, negation is not just considered to be a feature that "feeds" capital; it must also be understood politically, that is, in its autonomy as a class concept. One moment of negation, one type of Aufhebung, is elevated in an act of strong political partisanship. The perspective also explains Cacciari's later and perhaps most notorious comment—only one among a number to shock his comrades—when as a PCI parliamentary deputy he deployed negative thought to declare that what the working class needs is not Marx's theory of value but Nietzsche's will to power. The political act is emphasized over the passive object and objectification. "We have no models," Tronti claims; yet even when picking up Marx's metaphors—that the communist revolution would not need to wrap itself in togas he argues in his opening to the *Eighteenth Brumaire*—Tronti takes them in a decidedly Futurist direction:

> in June 1848 (that fateful month, a thousand times cursed by the bourgeoisie), and possibly even earlier, the working class took over the stage, and they have never left it since. In different periods they have voluntarily taken on different roles—as actors, as prompters, as technicians or stage-hands—while all the time waiting to wade into the theatre and attack the audience.[195]

From Marx to Marinetti—or, perhaps, "Marx plus Marinetti"—is not without its problems: the reduction of Marx's concept of revolutionary history to the futurists' aggressive version of *épater le bourgeois* might be one; Marinetti's political passage from social democrat to fascist another; a further might be the workerists' commitment to appropriating the gains of "negative thought" even in its Right-wing forms. We might want to explore the extent to which continuities can be traced between Left-Hegelianism (whether taking anarchist or communist directions) and an avant-garde *Weltanschauung* (the adoption of political forms, the manifestos, and language of intervention). Nevertheless, Tronti's account draws a vivid metaphor for varying forms of class presence on the "historical stage"; it also indicates the association of this type of radical politics with early avant-gardist techniques and their fetish of destruction.

The approaches taken, and terms developed, by Italian workerism were, and are, deserving of radical criticism. The famous Trontian inversion has a tendency to become political blindness; by way of "paradoxical" turns, political weaknesses can always be represented as strengths, defeats made over as victories. One commentator notes how out of the same interwar American struggles idealized in workerist theory came one of the triumphs of capitalist development: the destruction of American cities.[196] Workerism's "central thesis is false," another contends, but not without adding "alas!" The rejection of the problems of alienation and fetishism within workerism—a feature recognizable within Cacciari and Tafuri's critique of anxiety—leads to a one-sided conception of the worker as subject and a one-sided emphasis on political will. The emphasis on massification ignores the new forms of hierarchization emerging within both labor process and social relations at large. Tronti himself comes to take his distance from his own theses, insisting recently that *operaismo* should be seen as specifically tied to the practices of the Fordist and Taylorist workplace. (Negri, of course, now pushes further some of these problematic features: "multitude" may well claim to be beyond the mass worker—and even beyond its successor, the "social worker"—but there is a direct conceptual descendency; an emphasis on difference may have replaced the social homogenization characteristic of the mass worker, but the processual qualities of massification—now modified as leveling-in-multiplicity—are, along with the inversion thesis, still central to recent arguments.[197]) Nevertheless, for a while, the culture of this Italian Left had grounds for its sense of growing strength and wrought tangible gains. Despite serious flaws, the account produced an outlook—again for a period, although one considerably more sustained than "1968"—that disdained the type of disillusion and defeatism that elsewhere became so widespread. This is probably an important factor in workerism's resurgence among anticapitalist movements, as activists have sought to develop understandings that exceed both the fascination with *autonomism* (where the belief in the inversion was more exaggerated and yet less determinate) and the debilitating stream of constant reminders of capital's total domination.

Influential as the Frankfurt School was for these Italian intellectuals (especially for the insights it gave them into negative thought), social agency was firmly on the workerist agenda—and not just present for them as a historical fact of their time; to the point of being fetishized, it was the heartbeat of their outlook. They were not preoccupied by the iron cage of modernity, even when they advanced the idea of the "social factory." There was no need for placeholders under which the critical spirit might shelter, locations where nonidentity might hold out against the crushing identity logics of administered capitalism. There was little need either for utopia and anticipation (indeed, utopia was taken to have become an obstructive ideology). Echoing Panzieri's critique of the PCI and PSI, Tafuri rejected the claim among radical practitioners to be prefiguring revolutionary socialism. This is why he argued that there was no class architecture, only a class critique of architecture—a direct echo of some classic radical Bolshevik positions: Vološinov on "class language," Pashukanis on "class law," and Lenin and Trotsky's criticisms of Proletkult's class culture—and why Tafuri considered himself to be not a historian of architecture "but *also* a historian of architecture."[198] The role of the social subject was emphasized over that social structure—again, very different from other traditions from Western Marxism: the active agents of workerist theory were as far from the troubled selves of existentialism as they were from the Frankfurt School's near obliteration of the subject. (The gulf between Tafuri's approach and Jameson's—with the latter's predilection for "cultural logics" and structural necessities—should be apparent.)

The point of the politics represented by *Contropiano* was to break from the reformist and Keynesian logic of the established Left; to challenge the most sophisticated exponents of capitalist science; to recover Marx's concept of value as a class relationship; to conceive—and to realize—class as an independent subject. The disenchantment achieved in modern capitalist economic science needed to be matched, or outpaced, by the Left's. Remember: no more whining about capitalism's imbalances, no more nostalgia for past forms of life, but the counter assertion of proletarian interests. Tafuri's assessment of the avant-garde was based on the turn from an emphasis on capitalist development to that of this negative power. (His critique of negation was, additionally, directed at its complacent appropriation by the discipline of architecture and the latter's own adoption of avant-gardist ideology and myth.[199]) For intellectuals working from this perspective, historical analyses are intended to be foundations for a contemporary political strategy. Tafuri's judgments on the avant-garde are best understood, then, as being made much in the way that political militants might criticize the political mistakes of their forebears: the purpose is to understand the problems that beset the ambitions of avant-garde artists, the architects of the modern movement and those involved in Italy's postwar building. The effort is directed to reconsider issues of the past from the inside, to analyze moments in all their contradictions and to rethink architectural history as part of a history of social process. In their straight and unsentimental acknowledgement of disenchantment, and their proximity to negative thought, Tafuri judges some interwar avant-gardists to have been considerably ahead of the thinking of political theorists and activists (both of Left and Right). The Tafurian critique of the avant-garde is harsh, but it is certainly not posed in a dismissive tone or presented as proof of their inevitable assimilation or the all-powerfulness of postwar capital. Returning to the history of the emigration of western architects and planners to work on Soviet projects such as Magnitogorsk in an interview from the late nineteen-eighties, he argues the following:

> A crucial discovery lay in seeing that the principles of architecture elaborated in Weimar Republic Germany couldn't be applied in the context of the Soviet Union and the first Five Year Plan. But that's perceived as a defeat only by those who don't see

> the real historical consequences of what took place, and my criticisms [of postmodernism made in *The Sphere and the Labyrinth*] were directed against the voices that react to this with a funeral lament.[200]

The avant-garde is not just an object of study for which a workerist politics and philosophy of history provides an intellectual background: there already seems to be some closer affinity at work. In Tafuri's argument avant-gardists of the twenties are considered less as members of an artistic vanguard so much as seen as dramatizing the debates over negative thought. Indeed, he places some sections of the avant-garde as among the most advanced participants of negative thought, as agents making that thought into praxis, refusing to entertain tragic anxiety over modernity or over the loss of aura, and seeking neither consolation nor salvation. To this extent, as negative thinkers making their philosophy into praxis, such artists could be seen not simply as serving the needs of capitalist development in the period of Ford and Taylor but also as progenitors of workerism itself, claiming the disillusioned perspective appropriate to the Metropolis for progressive purposes—who, as we saw in the case of the Italian Futurists, abandon "all vindication of lost objects," or, as with El Lissitzky, embrace "devaluation" and "the universe without qualities," or, as with Schlemmer, both create and liberate the (political) subject as "marionette." Certainly, Tafuri separates out such early avant-gardists from postmodernism's half-hearted nihilism, which he sees as a consoling position that attempts to smooth over the effects of devaluation. The turn to action may well feed the needs of capital, but that does not mean that the system could simply claim ownership of that dynamic; that it claims it at all is but a reactive sign of action's origination. In this view, the stage of the Metropolis is subject to other motile forces and eruptions that threatened capital.

Tafuri's dialectic certainly has directionality; there is even the repetitive pattern of doubling, of contradictions, characteristic of *Theories and History of Architecture*, which owes much to the Wölfflinian mode of historical conceptualization in art. However, despite this, his history of the avant-garde is not orchestrated toward an inevitable telos nor is his critique intended to serve as a form of theoretical and historical closure. Like its object, this critique orients toward the future, and toward something like, to use Jameson's words, an "ontology of the present."[201] "Critical history" is not critique abstracted from history, but one to be inserted into "history as an event." And so, as Tafuri puts it in the introduction to *Theories and History*: "To criticise . . . means to catch the historical scent of phenomena, put them through the sieve of strict evaluation, show their mystifications, values, contradictions and internal dialectics, and explode their entire charge of meanings"; the point was to arrive at "the historical assessment of the present contradictions"; it is "to explain, to diagnose exactly and to avoid moralising in order to see, in the context of negative facts, what are the mistakes we are now paying for, and which are the new values nesting in the difficult and disconnected set of circumstances we live in, day by day."[202] This unashamed rhetoric of the Enlightenment—full of the language of diagnosis, open-mindedness and the "watchful eye" against (new) myths and prejudices—is one already painfully aware of the problems facing the very practice of criticism: its "complicity" with its object and the critical subject's placement in history. But the political project and its philosophy of history—most especially what to avoid—seems clear. "Desperate nihilism," Tafuri already writes in *Theories and History*, "belongs to those who, realising the wearing out of the myths at the base of their personal faith, can see in front of them only irrevocable *destiny*."[203] There is no given outcome to struggles; history has no preset destiny; and those committed to social liberation should entertain no desperation. Whatever the flaws of this body of intellectual work, it remained more resilient than most others to the practice of historical pessimism even as it actively engaged with the historical pessimism of negative thought.

Tafuri's represents one of the most explicit and extended articulations on negation and art to emerge from the New Left, not to mention a specific and distinctive political expression of the philosophical problem of Left Hegelianism and nihilism. Its full implications are still to be reckoned with, and, perhaps, still to be played out. Between the methodological positivism and (sometimes blind) political optimism of *operaismo*, on the one hand, and, on the other, the historical pessimism generated out of the Left-Hegelian and Western Marxism traditions—between militant autonomy and the melancholy of alienation, or between the emphasis on politics and on economic theory—there is still much to be negotiated. Advanced debates over the critical status of the avant-garde, and of radical cultural practice more generally, live out the tensions between intellectual and political orientations such as these. More than many in the workerist movements—and, indeed, the PCI—Tafuri and Cacciari explicitly engaged the antipositivistic arguments deriving from German philosophy and Western Marxism; more than the Frankfurt School philosophers, their thinking was shaped by a culture of militant praxis. Ironically, the claim that Tafuri's account is pessimistic speaks more to the bleakness of the intervening neoliberal years than it does to Tafuri's approach to avant-gardism and negation. This argument seemed little more than an interesting footnote until late summer 2008 when the global banking crisis brought debates over Keynesianism back onto the political and economic stage.

Chapter 3

Absolute Dialectical Unrest

Or, the Dizziness of a Perpetually Self-Engendered Disorder

> The mistake consists in taking such abstract forms as "the same" and "not the same," "identity" and "non-identity" to be something true, fixed and actual, and in resting on them. Neither the one nor the other has truth; the truth is just their movement in which simple sameness is an abstraction and hence absolute difference, but this, as difference in itself, is distinguished from itself and is therefore selfsameness.
>
> —G. W. F. Hegel[1]

Part I

At the beginning of the nineteen-eighties, theorists in art detected a new "allegorical impulse."[2] Taking their distance from what they perceived to be modernism's "symbolic" practices, these writers set out to reverse a longstanding hierarchy in which the values associated with the symbol (essence, aesthetic immediacy, and organic plenitude) were privileged over those of allegory (mediacy and mechanical re-presentation). In the visual arts—and especially among writers closely identified with the journal *October*—this revived interest in the allegorical mode fused the critique of Clement Greenberg with a growing interest in poststructuralist thought. It also drew on the interest in modernism's radical manifestations in the historical avant-garde (as well as ideas from Brecht and Benjamin), which had been recovered and developed by artists and theorists during the nineteen-seventies, and which had been excluded from the Greenbergian canon. Emerging under the aegis of Rosalind Krauss, one of the late twentieth century's most significant theorists of art and a founding editor of *October*, this new account became particularly influential.[3] In its dissemination, the argument soon lost much of the critical impulse that had animated its earlier formulations, frequently collapsing into a binarism where postmodernism and allegory figured as critical terms in opposition to (what was taken to be) the ideologically normative conditions of modernism and the symbol. Without doubt, this schema had the virtue of clarity in the pedagogic context, but it came at a price. By the nineties it was clear that a number of the key emphases originally promoted under the heading of "postmodern allegory," such as "ambiguity" and "undecidability," had undergone a paradoxical inversion; sometimes they had even been recuperated, uncritically, into a loose symbolic aesthetic of the ineffable nature of art. While certainly not guilty of such gross debasements, the instigators of allegory's revival in art nevertheless contributed to the work of aesthetic schematization. These thinkers were instrumental in revealing the conceptual assumptions undergirding much of the dominant formalist art theory, but their role was also to establish and fix a particular value scheme in which "inorganic" conceptions were privileged over "organic" ones and "fragments" were set against "totality." All too often, dialectical thinking proved the victim of this realignment; sometimes it was the explicit target, accused of surreptitiously purveying the symbol's organicist prejudices.

The claims for a postmodern "allegorical impulse," formulated around 1980, provide a particularly interesting example of the wider poststructuralist abandonment of dialectic modes of thought. The interest in allegory in the context of the visual and plastic arts followed on a similar, albeit earlier, shift in the field of literary history and theory.[4] In art theory the tenor of the debate altered, in part echoing changes within literature's own theoretical debates that occurred at this time; what began in literature as an attempt to reevaluate allegory in a more sympathetic light and to draw a long overlooked corpus more centrally into the literary canon hardened into a more aggressive polemic within the visual arts. Even the much stronger claim that we had "reentered an allegorical age," which originated in literary theory, sounded a different register.[5] Allegory was not merely defended as an alternative aesthetic device; drawing on studies by Benjamin and more recent writings by the literary theorist Paul de Man, allegory was positioned as a new, or reclaimed, worldview around which a politics of aesthetic practice was staked.

The demise of allegory's promise in previous ages is generally identified with the growing suspicion toward rhetoric; similarly, the renewed interest in allegory in the late twentieth century is closely allied to the linguistic turn.[6] Modernist critics were challenged for having contained discussion of the visual arts within "a realm of exclusive visuality," for emphasizing "opticality," and for deeming irrelevant the role of narrative.[7] *October* provided a key forum for exploring and promoting the "eruption of language into the aesthetic field."[8] For Craig Owens, this linguistic upsurge represented the return of modernism's repressed. Around 1980, Owens in particular—but also Douglas Crimp and, with some qualifications, Benjamin H. D. Buchloh—contributed to this turn to allegory, where it became a key component in theorizing a radical postmodern project.[9] In the context where "the public sphere was contracting," the journal set out to provide, as Krauss put it, a vehicle for "liberal and left-wing" values.[10] Influenced by journals such as *Tel Quel*, and drawing on French structuralism and poststructuralism, the writers associated with *October* mobilized Foucault and Lacan to challenge the separation of word and image defended by mainstream modernist aesthetics. Foucault's 1968 essay "*Ceci n'est pas une pipe*" was specifically translated for *October*'s first issue in 1976; indeed, it was *Artforum*'s refusal to publish this piece that had prompted the resignation of Krauss and Annette Michelson from *Artforum*'s board and led to the establishment of *October*. Not that *Artforum* was, at this point, the bastion of modernism and formalism: Krauss and Michelson objected to the increasing politicization of its editorship under John Coplans and Max Kozloff, believing its conception of the relation of art and politics to be too reductive. *Artforum*, they thought, was insufficiently open to intellectual innovation and they looked to European Critical Theory to provide fresh ways "to characterize new departures."[11] The political framing of the birth of *October*, then, was rather more nuanced than Krauss's claim to liberal and Left-wing values might suggest; these values needed to be pursued in a particular intellectual direction. Similarly, the journal did not simply promote linguistic models as such but specifically opposed referential conceptions of language. Initially focusing on recovering the legacies of the European radical avant-garde movements of the twenties and thirties, and on introducing theoretical categories associated with recent continental philosophy, *October* soon became identified as "the journal that was most central in theorizing postmodernism in the visual arts."[12] The account that became prominent in art theory at this time frequently implied that allegory should be taken not only as a critique of the symbol but also of the idea of the "real." Thus, postmodern critics tended to run together realism with the symbol; and, the terms of their promotion of allegory can also be glimpsed in the "critique of representation."

The attention to allegory forms a particularly significant aspect of *October*'s critical orientation. It is Craig Owens's now-canonical two-part essay of 1980, "The Allegorical Impulse: Towards a Theory of Postmodernism," that provides the most programmatic statement on the subject of postmodernism and allegory.[13] As Owens suggests, with reference to Walter Benjamin: "Allegory is consistently attracted to the fragmentary, the imperfect, the incomplete."[14] These characteristics figure prominently in the work of those artists who openly challenged the high-modernist protocols that had been established in the United States by the sixties; attracted to the social and hybrid qualities of the interwar avant-gardes, these artists were drawn to the very projects that Greenberg ignored and disliked. Developing from certain strains of conceptual art, this new work can be best typified by art that appropriates social signifiers or accumulates cultural objects; practices that engage with the specificities of particular spaces and institutional contexts; and work that hybridizes media and representational modes, emphasizing art's "pictogrammatical" engagement in the wider field of metatextuality and social discursivity. The aesthetic claims to autonomy and transcendence associated with modernism and those offering directness of expressive effect, "pure presence" and "essence," are contrasted to artistic practices that emphasize contingency, interrupt conventional expectations, and frustrate the desires associated with aesthetic pleasure.[15] Owens shapes his account of allegory with resources drawn from structuralism and poststructuralism, foregrounding the supplement, the palimpsest, and the shift from speech to writing or from image to text, and attending to counternarrative tendencies, the deferral of meaning, and the breaking of sign from referent and signifier from signified. In identifying the return of allegory with challenges to "the symbolic, totalizing impulse which characterizes modernist art," Owens focuses specifically on the self-conscious employment of photography used by Robert Longo, Sherrie Levine, and Cindy Sherman.[16] Indeed, he designates photography as an allegorical art par excellence.[17] The mechanical nature of its reproducibility, which had traditionally ensured its lowly status as art, was—from an allegorical perspective—precisely the source of photography's strength. Levine's rephotographing of canonical images, such as those by Ed Weston or Ansel Adams, are taken as explicit displays of the device of appropriation, which treat its "(re)sources" as textual; for Owens, this both acknowledges and heightens our awareness of photography's status as a discursive practice. He notes how, as an earlier example of site specificity, Robert Smithson's *Spiral Jetty* (1970) demonstrated the impermanence appropriate to allegorical art; as it decays and dissolves back into the landscape, the work is preserved only by the photographic record, which in turn stands as an allegorical ruin.

Krauss also gives considerable attention to the subject of mechanical reproduction: the first issue of *October* includes her essay on video; her influential work on the index appeared in 1977; two years later, her account of Alfred Stieglitz's images specifically highlights—and undoes—the modernist-symbolic claims of his practice.[18] The fifth issue of *October*, published in summer 1978, is entirely devoted to photography, indicating its importance for the journal's project; both Owens and Crimp contributed to this issue. Owens's essay—and this is also an important feature of Krauss's articles—is at pains to displace photographic imagery from its traditional association with transparency and the presence of the objects or subjects being photographed. Photography is identified with textuality and with Jacques Derrida's description of the textual operations of *mise en abyme*. Owens tracks "the indefinite play of substitution, repetition, the splitting of the self"—the internal analogues, mirroring, and reduplication—at play in the images of Brassaï, Lady Clementina Hawarden, Walker Evans, and Robert Smithson.[19] Here photography figures as a self-referential system of signification, and Owens focuses on examples that can be seen to foreground their own process. "Photographs," he argues, "are but one link in a potentially endless chain of reduplication"; "if Evans's photograph depicts a reality outside of the photograph, that reality is nonetheless *wholly conditioned by* the properties of the image." Shifting his attention to Smithson, he underscores the same point: "the hierarchy of object and representation . . . is collapsed" and the "representation can no longer be grounded . . . in presence."[20]

Douglas Crimp's essay, which derived from his doctoral work under Krauss, addresses the photographic work of Edgar Degas, pursuing the more conventional art-historical task of reevaluating a specific body of hitherto overlooked work, in this case, to prove that the artist was a serious, rather than amateur, photographer.[21] However, in a series of articles written in the early eighties, Crimp develops his interest in a much more ambitious direction, where he elaborates arguments about the medium of photography and its impact on archives. He argues that with the incorporation of photography into the canon of art, and its subjection to the homogenizing museological gaze, a disruptive impulse was unleashed. Crimp describes Andre Malraux's "Museum without Walls," a museum achieved through the possibilities provided by mass-circulating photographic reproductions, which, Malraux believes, can act as a universal language, enabling the free comparison the cultural objects and artworks housed in collections around the globe. Malraux's museum without walls contains photographs of art objects, but also photographs as entities in their own right, and it is the latter category that, Crimp argues, threatens Malraux's entire project. While Malraux treats photography as a recording device, he is able to contain its threat, but once he includes photographic objects in his museum the destructive potential of photography's uncontainable heterogeneity is set loose. "Even photography cannot hypostatize style from a photograph," Crimp observes.[22] Appearing at the heart of Malraux's universalizing project, this force unravels his system of knowledge. The same destructive force—which Crimp sees as inherent to the medium of photography—compromises the efforts of John Szarkowski, the Museum of Modern Art's influential former curator of photography, to institutionalize a canon of photographic art. Crimp understands Szarkowski's role at MoMA as an attempt to override photography's plurality and submit it to a modernist aesthetic; however, in doing so, Crimp argues, Szarkowski is forced to revise the modernist paradigm itself:

> and it can happen only because that paradigm has indeed become dysfunctional . . . . It is photography's reevaluation as a modernist medium that signals the end of modernism. Postmodernism begins when photography comes to pervert modernism.[23]

In yet another essay, which highlights the work of Sherman and Prince, Crimp claims that postmodern photography works "in complicity with" the traditions of modern art photography, but that "it does so only in order to subvert and exceed them."[24]

While emerging poststructuralist theories increasingly distanced themselves from dialectical thought, Benjamin Buchloh remained very sympathetic, drawing on arguments outlined in Adorno and Horkheimer's critique of the culture industry. More cautious than either Crimp or Owens (and indeed, than Krauss and Hal Foster) with regard to the critical claims of postmodernism, his account focuses on an alternative allegorical current *within* modernity's avant-garde. For Buchloh, the critique of Greenberg's modernism involves, in part, a reappraisal of the interwar avant-gardes and their postwar neo-Duchampian progeny. In his essay "Allegorical Procedures" (1982), Buchloh follows Benjamin and links allegory to the European avant-gardes's "dialectic of montage."[25] "The procedure of montage," he argues, "is one in which all allegorical principles are executed: appropriation and depletion of meaning, fragmentation and dialectical juxtaposition of fragments, and separation of signifier and signified."[26] While sometimes adopting similar techniques, he sees the postwar neo-avant-garde, in contrast, as falling "for the premature delusion of an immediate reconciliation between high art and mass culture"; as succumbing, in other words, to a "liberal reconciliation."[27] Buchloh suggests that the ground for a return to allegorical procedures proper was prepared in the midsixties by artists such as Michael Asher and Marcel Broodthaers, because, as he puts it, their artwork "integrates the historical ramifications of the Ready-made model and the consequences of a self-referential analysis of the pictorial construction itself."[28] Entailing both recognition of, and intervention into, the institutional framework, their work engages with the processes of signification.[29] Thus, he sees these artists operating at two levels, their work being simultaneously a critique of the institution of art and a critique of the sign. This project of "allegorical deconstruction"—which, Buchloh reminds us, both negates and does not negate—paves the way for a paradigmatic shift during the late seventies, when artists such as Dara Birnbaum, Louise Lawler, Martha Rosler, and Sherrie Levine develop new strategies for sustaining these critical projects.[30] Buchloh states explicitly that these developments produce a situation that, like Cubism earlier in the century, can be seen as initiating a new "historical logic," the ramifications of which cannot be retracted. There has been, he concludes, "an irreversible change in the cognitive conditions of art production."[31]

The paradigmatic shift of the late nineteen-seventies, to which Buchloh refers, alerts us to the emergence of a new generation of artists. Names that are now firmly identified in the histories of late-twentieth-century art with postmodernism—Barbara Kruger, Sherrie Levine, Troy Brauntuch, Alan McCollum, Robert Longo, Jenny Holzer, Richard Prince, Cindy Sherman, Louise Lawler—came to prominence at this time. The 1977 group exhibition entitled *Pictures*, held at New

York's Artists Space in the autumn—which, in addition to work by Levine, Longo, and Brauntuch, included work by Jack Goldstein and Philip Smith—is often taken as foundational for this tendency. Deliberately transmedial in conception, *Pictures* involved "recognizable images," but also took its title as a verb to refer "to a mental process as well as the production of an aesthetic object."[32] Organized by Crimp, the catalogue text was expanded in an article published in *October* a year and a half later, and *Pictures* not only came to stand for those artists included in the exhibition but also encompassed a generation of conceptually informed and culturally aware practitioners who deployed strategies of representational replication. Hal Foster provides a classic articulation of this mode in his essay "Subversive Signs" of 1982, describing the artist as "a manipulator of signs rather than a producer of art objects," and the viewer as "an active reader of messages rather than a passive contemplator of the aesthetic or consumer of the spectacular."[33] Aware of their "entanglement in discourse," these artists can be seen as a counter to the neoexpressionistic revival that fueled an expanding corporate art market.[34] Describing them as the "legatees of conceptual art," Foster points to their interrogations of subject identity, social constructions of gender and sexuality, regimes of desire, and image fetishism, which he sees as signaling their opposition to the neoconservative turn in world politics, or to what might now be seen as the formative period of neoliberalism.[35]

The terms of the allegory-symbol distinction, then, served the challenge to Greenberg's version of modernism. Owens specifically emphasizes his target to be not modernist practice but rather modernist theory; modernist texts, he argues, "contain within themselves the seed of their own allegorization," and he concurs with Benjamin that there is "an allegorical impulse at the origin of modernism in the arts."[36] According to Owens, the recent revival of this impulse displaced the dominance by the symbolic mode, which, for the two centuries from Romanticism through modernism, had subjected allegory to, what he calls, a "permanent strategy" of suppression.[37] Walter Benjamin is no less forthright in his judgment: we should understand the reign of the symbol, which he dates from around 1800, as "the tyranny of a usurper."[38] The symbolic claims of Romantic and post-Romantic thought, Benjamin writes, are tantamount to "illegitimate talk" and "abuse," which staged allegory as the symbol's mere "speculative counterpart."[39] Exactly how to date the emergence of the symbol's dominance is a matter of some debate. The literary historian Edwin Honig blames post-Romantic thought for a programmatic denial of the subject-object split, believing the Romantics still held to an allegorical worldview; Paul de Man draws the line within the Romantics, distinguishing an earlier allegorical formation from a later adherence to the symbol. Benjamin himself admits that the decline of the allegorical genre was due as much to the weakness of its supporters as to the sustained attacks mounted by its detractors.[40]

Whenever the exact moment of this suppression occurred—with, within, or after the Romantics—there is a broad consensus that the dominant "aesthetic ideology" of the symbol, to use de Man's formulation, thereafter casts allegory in a value-laden pattern of figuration: as the prosaic term to the symbol's poetry; as the mechanical concept to the organic idea; as the conventional sign to the essential and true symbol.[41] Treated as a lesser mode of representation—precisely as a *re*-presentation—allegory is compared unfavorably to the symbol's claim to *present*. In contrast to the symbol's privileging of immediacy, presence, and identity (the term derives from *sum-ballein*, to "throw together"), allegory's nonidentity is restricted to the disjunctive distancing of "other-speaking" (*allos-agoreuein*). For Ernst Bloch, the "strict Absolute or final sense" of *Unitas* contrasts with the transitory and processual becoming of allegory's *Alteritas*.[42] Symbolic language, Benjamin writes, insists "on the indivisible unity of form and content"; it is "an unbroken whole" of the beautiful and the divine, the "unity of the material and the transcendent object."[43] The symbolic mode posits an "intimate unity" between subjective experience and its representation, and, as de Man argues, between "the image that rises up before the senses and the supersensory totality that the image suggests"; it "postulates the possibility of identity or identification."[44] Where the symbol is venerated as concentrated Being—a total and direct incarnation of the idea, beautiful, effective, fertile, unique, and transcendent—allegory is characterized as didactic, ugly, ineffective, and barren, as a sign with finite meanings. Unable to rival the symbol's "translucence," allegory is thus dismissed by Samuel Taylor Coleridge as "a translation of abstract notions into a picture-language, which is itself nothing but an abstraction from objects of the senses":

> It is among the miseries of the present age that it recognizes no medium between *Literal* and *Metaphorical*. Faith is either to be buried in the dead letter, or its name and honors usurped by a

> counterfeit product of the mechanical understanding, which in the blindness of self-complacency confounds SYMBOLS with ALLEGORIES.[45]

To the symbolic mind, allegory seems artificial, excessive, mechanical, inorganic, obvious, sermon-like, doctrinal, and dated, comprising formulaic plots with undeveloped characters who are simply ciphers for the delivery of didactic messages. Such characteristics come close to those employed by Clement Greenberg in the late nineteen-sixties when he dismisses those modes of contemporary art that, in his opinion, depend on "a feat of ideation" or that aspire to "the look of non-art."[46] Michael Fried identifies a similar problem in his critique of Robert Morris, Donald Judd, and Tony Smith, where he accuses them of being "literalists." When, in "Art and Objecthood" (1967), he contrasts the contingent and anthropomorphic "presence" of "literalism" with the "presentness" of true art—concluding that "presentness is grace"—he employs a judgment that privileges the immanence associated with the symbolic impulse over its temporally specific opposite.[47] As Benjamin might have pointed out—citing the terms set out by Friedrich Creuzer—Fried establishes his contrast between art and objecthood as that between the former's "momentariness" (the "mystical instant" or "momentary totality") and literalism's aesthetic experience as a "progression in a series of moments."[48]

Allegorical Disjunction

Despite a longstanding suspicion and antipathy toward allegory, usually traced back to Plato, we should not overlook its significance as an aesthetic genre. From a rhetorical perspective, allegory is an "extended trope" or "figure," which can incorporate other tropes.[49] Allegory is a "twice-told tale," which depends upon an original textual authority.[50] Its historical roots seem to lie in textual exegesis, or allegoresis, where readers trace correspondences between the "literal" and "figural" (or "symbolic") levels of a text: we can see this with the interpretations of the Odyssey and Iliad, and with the Pergamean school's defense of Homer from the Pre-Socratics, in Talmudic scholarship, or in Biblical analyses tracing analogies between Old and New Testament themes (the latter largely driven, Maureen Quilligan informs us, by the need to make the ancient scriptures meaningful for a changing world).[51] In its typical literary form, an allegorical text is structured around the theme of a journey or quest (as in, for example, Dante's *Commedia*, John Bunyan's *Pilgrim's Progress*, Edmund Spenser's *The Faerie Queene*). These journeys characteristically involve a separation from some first place and moment, an alienation that is necessary in order for the protagonist to face and overcome a number of challenges. Progressing via battles and psychomachic struggles, the main character encounters a range of personifications or daemons (some good, some bad; sometimes entirely other to his being, sometimes revealed as aspects of himself). These travels can be seen as allegories of (Christian) spiritual self-discovery, leading to a more profound reconciliation with God, the "home" that is finally regained—a concluding point that is explicitly denied in *modern* versions of allegory.

A great deal turns on allegory's diremptive or postlapsarian nature. Angus Fletcher points to the themes of distance, remoteness, and alienation: allegory, he notes, involves a "characteristic splitting."[52] Edwin Honig describes allegory as "the product of disjunction between the workings of reason and the workings of the imagination" and "a disjunction between the 'inside' and the 'outside.'"[53] For Paul de Man, allegory is structured in such a way that:

> the relationship between sign and meaning is discontinuous, involving an *extraneous* principle that determines the point and the manner at and in which the relationship is articulated . . . . [T]he sign points to something that differs from its literal meaning and has for its function the thematization of this difference.[54]

Returning to the subject on several occasions, Benjamin also draws attention to the alienating characteristics of the allegorical mode. As the analogy with montage indicates, Benjamin emphasizes allegory's dissociative method: "the disjunctive, atomizing principle of the allegorical approach."[55]

Pointing to radical nonidentity and disjunction, and explicitly challenging the symbol's characteristic qualities of identity, unity, and immediacy, the appeal of the allegorical mode to a burgeoning postmodern theory appears to be clear. Crimp explicitly acknowledges the Foucauldian basis of "concepts such as discontinuity, rupture, threshold, limit, and transformation," articulated in *The Archeology of Knowledge*.[56] (We should not forget the centrality of history as "event," which was encountered in the previous chapter, to Foucault's study.) Within this Foucauldian frame, Crimp draws on Eugenio Donato's analysis of Gustave Flaubert's *Bouvard et Pécuchet*.[57] Returning us to the terms directly associated with allegory, Donato notes how the fiction of coherence offered by the various systems of knowledge employed by Bouvard and Pécuchet dissolves into "'bric-a-brac,' a heap of meaningless and valueless fragments of objects which are incapable of substituting themselves either metonymically for the original objects or metaphorically for their representations."[58] For Owens the model of this allegorical unraveling and dissolution takes a more deconstructive direction and acknowledges that there is a constitutive "complicity" between the deconstructive process and the object of deconstruction.[59] His argument aims, he asserts, neither to "bracket" nor to "suspend" the referent but to "problematize the activity of reference."[60] This important emphasis in Owens's conclusion, however, becomes submerged under the broader themes of the postmodern debate: the critique of originality, authorship, expression, and authenticity; the critique of origins and presence, of totalization and closure. Interrogating the activity of reference is at the core of the symbol-allegory discussion, but when swept into the relativist conclusions of art-theoretical discourse in the eighties, Owens's insight was itself reduced to a closed and totalized figure.

As should already be apparent, the claims made for, or against, allegory are rarely modest. Helga Geyer-Ryan considers allegory to be a patriarchal repression of maternal corporeality, where the female body is made into a sign for masculine meaning. As she remarks of Benjamin's account (to which she is not sympathetic): "allegory is a battlefield on which the body—especially the female body—is torn apart."[61] Christine Buci-Glucksmann makes very different feminist claims on the basis of the same material:

> the metaphor of the feminine then rises up as an element in the break with a certain discredited rationality based upon the idea of a historical and symbolic continuum. It does this by designating a new heterogeneity, a new otherness . . . . a whole network of negativity escaping the dominance of the Concept, a whole "culture" of the feminine . . . . a culture of *Spaltung*, of splitting . . . .[62]

For Buci-Glucksmann, allegorical or "baroque reason" is the reason of the Other; it is a "consciousness of rupture" and negation, which she further identifies with a Judaic sense of history, as radically unstable, insecure, and catastrophic.[63] Allegory, for Buci-Glucksmann, recognizes a "gulf between reality and illusion" that "cannot be bridged."[64] Geyer-Ryan, for her part, opposes both the Saussurean conception of the sign and Benjaminian allegory for positing an "unmediated construction"—an "abyss" between signifier and signified "bridged only by conventions," and in which "violence reveals itself as a strain on meaning."[65] As we have already seen in earlier chapters, the terms of this discussion pertain to political debates over dialectics, negation, and mediation, especially as developed among theorists of the New Left. Buci-Glucksmann explicitly associates allegory's "reason of an unreconciled history" not with the "Marxisms of progress" but with what she calls "a rent Marxism."[66]

The Reification of Allegory

There is little doubt that the two major theorists of allegory, Walter Benjamin and Paul de Man, find in the symbol immediacy, presence, identity, and transcendence, and that they emphasize allegory's qualities of nonidentity, rupture, disjunction, distance, and fragmentation. However, I shall argue that what matters for both are, firstly, the dynamic relations within the conceptualization of allegory itself and, secondly, the ways that these two aesthetic categories are distinguished from one another. The stakes in the allegory-symbol debate are perceived as high ones. For de Man, the issue has consequences for our understanding of a number of traditional philosophical oppositions: mind and nature, subject and object, depth and surface. He specifically opposes the aesthetic ideology engendered by Romanticism, arguing that, over the course of the eighteenth century, the method of "associative analogy"—encompassing, for instance, analogies between self and nonself, subject and object, consciousness and nature—gives way to a more vital form of analogy, which is characterized by "affinity" and "sympathy." Alterations in the main topos of subjectivity underpin this change: the earlier emphasis on the intersubjective passes over to that of the intrasubjective, and the subject emerges as the "infinite self."[67] Among de Man's specific targets are the tropes of symbol, metaphor, synecdoche, mimesis, and personification—that is, categories that, he thinks, privilege immediacy, presence, plenitude, and genetic (or organic) causality. De Man suggests that the tropes and figures of aesthetic ideology are not only the *objects* of literary criticism and theory (those found in the texts under study); they also structure the patterns of conceptualization employed by critics and theorists. Thus, for him, a trope such as synecdoche shapes perception, cognition, and understanding.[68] Whereas metonymy can only establish a causal link between part and whole, synecdoche draws an "intimate unity."[69] "The aesthetic," he writes, "is, by definition, a seductive notion that appeals to the pleasure principle, a eudaemonic judgment that can displace and conceal values of truth and falsehood likely to be more resilient to desire than values of pleasure and pain."[70] De Man conceives the aesthetic as containing dangerous lures that lie in wait for uncritical readers, those who would evade confronting the materiality of texts and their status as tropological events. This strong antipathy to aesthetic ideology, it has been argued, may be rooted in de Man's recoil from—and compensation for—his wartime involvement, as a columnist in Nazi-occupied Belgium, with the collaborationist newspaper *Le Soir*.[71] Without minimizing the anti-Semitism expressed in his early journalism, de Man's later position on aesthetic ideology seems to mark a distance from the false reconciliations of fascist aesthetics with the emotive manipulations of rhetorical effects. This rejection of aesthetic ideology concurs, in certain respects, with Benjamin's critique of the "aestheticization of politics," but where Benjamin calls for the "politicization of art," de Man delves into the intricacies of tropic movements, probing their implicit claims in order to catch sleights of hand in the most deep-seated and seemingly naturalized modes of human rhetoric.

De Man not only criticizes symbolic thought as the "temptation of immediacy," he goes further and suggests that this way of thinking produces a premature synthesis of contradictions.[72] In particular, his target is the sublation (the Hegelian *Aufhebung*) associated with dialectical processes of thought. Thus, it would be a mistake to understand de Man's resistance to the symbol just as objections to simple immediacy and pure presence. Such a conception would identify the symbol with no more than the first moment of the dialectic. De Man acknowledges that the more thoughtful literary and cultural critics—he names Stephen Ullmann, Peter Szondi, and Jean Starobinski—do not conceive of the symbol in this way, but instead view it as a form of dialectical synthesis. De Man, unlike some other commentators, therefore distinguishes these two understandings of the symbol; of these, those predicated on the Aufhebung, he thinks, represent a far stronger interpretive claim than those premised on simple immediacy. Thus, he challenges Ullmann's efforts to draw continuity between surface and depth, between sensory and subjective experience;[73] he argues against William Empson's characterization of art as the reconciliation of opposites;[74] he criticizes Starobinski's conception of art as a mediation of spirit and world and as a way beyond alienation; and counters Szondi's argument for "a reconciliation between the ideal and the real as the result of an action or the activity of the mind."[75] The dangers of "impatient 'pastoral' thought: formalism, false historicism, and utopianism"; the will to reconcile and desire to redeem through art: against all these one must be constantly vigilant, for, in de Man's view, these are the traps of aesthetic ideology.[76]

Both Benjamin and de Man emphasize allegory's temporal character.[77] In advocating a "truly historical poetics," de Man suggests that the essential point about the allegorical mode is its "truly temporal dimensions."[78] Allegory represents a "repetition . . . of a previous sign with which it can never coincide," and it "renounc[es] the nostalgia and the desire to coincide, it establishes its language in the void of this temporal difference."[79] This "unveiling of an authentically temporal destiny," or a "discovery of a truly temporal predicament," represents the unmasking of human self-mystification, specifically the attempt to seek immutability through an "affinity" with nature.[80]

> The dialectical relationship between subject and object is no longer the central statement of romantic thought, but this dialectic is now located entirely in the temporal relationships that exist within a system of allegorical signs. It becomes a conflict between a conception of the self seen in its authentically temporal predicament and a defensive strategy that tries to hide from this negative self knowledge.[81]

This is much like Benjamin's figure of the death's-head, reminding us of our mutability, not simply as a dent of *vanitas* to human pride but more specifically as a device to hinder any attempt to hide from this knowledge. The issue, then, turns on challenging those modes of thinking that try to sustain the illusion and refuse to face up to the truth. This strong claim to truth is only one of the ways that de Man and Benjamin differ fundamentally from Owens and Crimp.

Intervening in the *October* debates in the early eighties, Stephen Melville was quick to identify a number of problems with the accounts of allegory's reemergence. Drawing on deconstructive philosophy, Melville is critical of the *October* project, which he refers to as "this baggy monster," a position emerging from a self-referring community of interests whose participants are characterized by an "embeddedness in one another."[82] His main criticism is that *October* posits a "radical distinction and exclusion" between modernism and postmodernism, which he sees as a belated echo of Michael Fried's efforts to draw a sharp line between "art" and "objecthood." However, Melville suggests, Owens fails to recognize the extent to which, in Fried's hands, this radical distinction unraveled to the point of "break down."[83] As Melville puts it:

> Where postmodernism would mean something more radically separate from modernism (where it would forget modernism) it will end by forgetting itself as well—and it will do so by falling into the trap of modernism's favored mode of (self-)forgetting, the (non)dialectic of the mere and the pure. Postmodernism would proof itself against this risk by appealing to a deconstructive impulse working beyond the (merely) self-critical tendency of modernism—but the risk shows itself in the bare statement: the practice of deconstruction cannot rigorously hold itself apart from something called self-criticism except by hypostatizing its self in just the way it would avoid; it can articulate itself only insofar as it acknowledges explicitly its emergence from and dependence upon what might otherwise appear as mere self-criticism.[84]

For Melville, then, Owens's account represents a deallegorization of allegory and he argues, in contrast, that the postmodern can only be properly understood as "itself an allegory of the modern." In another critique, Fred Orton also questions the way the postmodern-modern distinction was hypostatized through the opposition of allegory and symbol.[85] In his study of Jasper Johns's *Flag*, of 1954–1955, he notes how the work articulates "well-rehearsed" oppositions—such as abstraction and representation, surface and subject—but also disarticulates them.[86] Allegorical disjunction is a central issue in this study, but Orton is concerned with the relational character, its "non-synthesized tension" or aporia.[87] Also drawing on deconstruction, Orton filters the distinction through a self-reflexive form of materialism, and while the privileging of subject matter at the expense of form is commonly perceived as a weakness of allegory, Orton refuses to deprioritize "surface matter" in his allegorical reading of Johns.

Allegorical criticism often views the literal level of the text simply as a code to be cracked, so that its hidden meanings can be revealed. Strictly speaking, this is the practice of allegoresis rather than allegory—a distinction explored by Maureen Quilligan. She argues that the *allos* of allegory refers not to some underlying meaning but rather to the polysemic capacity of language. Quilligan sees allegories as "narrative investigations of their own threshold texts" which typically appear in the opening passages.[88] In this sense, the act of interpretation is understood to be consciously internalized within the literary form as a mode of self-referential allegoresis. Thus allegorical texts are considerations of the potential of language to lead the reader astray with allegory's "dangerous polysemousness."[89] Angus Fletcher also argues that the purpose of allegory is to put secondary meanings into play, to induce dualities, pluralities, and polysemy.[90] For Honig, the analogies between its "realistic" (or better, its "literal") and its "symbolic" (or "figural") levels produce, when well handled, an "allegorical waver."[91] The ambivalence generated oscillates not only between different meanings but between truth and falsehood, or between good and evil.[92] Allegory can thus be seen as double or duplicitous. "Allegory's natural theme is temptation," noted C. S. Lewis; and another commentator pointed out that "Satan is the father of lies, the father of Sin, and in a special sense, the father of allegory."[93] Indeed, the key allegorical agent in Fletcher's account, the daemon, probably has at its etymological root the idea of division.[94] Nevertheless, in her discussion of allegorical literature, Quilligan explicitly rejects the claims to allegorical disjunction. Instead of attending to mere disjunctions of words and meaning, of literal and metaphorical reading, or to the gaps between different levels of meaning, she argues that we should be concerned with "the relationships across the gaps," tensions registering through the very literalness of the text.[95] In contrast to the "vertical" interpretation of allegoresis, allegories involve complex webs of meaning spun across "horizontal verbal surfaces."[96] Disjunctions and juxtapositions encountered on these surfaces, then, must be treated with caution. They tempt the reader to make translations in an act of resolution.[97] Allegorical texts aim, she suggests, to engage the reader in acts of critical self-reflexivity: "the text constantly invites and then exposes the reader's imposition of meaning," so that interpretations must be reinterpreted and their validity constantly reconsidered.[98]

Quilligan, Orton, and Melville all draw on deconstructive analysis, although they differ in their emphasis and concerns. All three writers direct our attention to the literality of texts or artworks. As Quilligan demonstrates, the reader is invited to live out the dilemmas of *allos* in relation to a material text—as a practice of engagement both with the literary text, and also with the process of critical reflexivity itself. What is offered by these accounts is an understanding of allegory that goes beyond the routine recognition of allegory's inherent disjunctiveness. Indeed, these theorists draw attention to the problem of hypostatizing the very opposition of allegory and symbol. Thus, close rhetorical attention opens onto questions of process and practice, and onto important ethical and epistemological dilemmas and decisions. Similarly, to be aware of the danger of hypostatization of allegory's disjunctive qualities is to be aware of the threat to the motility required of allegorical consciousness and its historical self-awareness. The emphasis on temporality that is prominent in discussions of allegory has to be lived out not just as a counterpoint to the "eternalist schemata" of the symbol, but, crucially, in relation to allegory itself.[99] We must pursue, therefore, the motile and temporal dimensions of allegorical work by looking more carefully at the transitive qualities at work within its disjunctive operations.

Part II

The writings of Walter Benjamin and Paul de Man provide the standard reference points for postmodern accounts of allegory, and it will be important to explore a different perspective on allegory and symbol via an extended reading of the work of these key authors. We need to develop a rather different framing of allegory from the one that prevails. Treating the allegory-symbol distinction as a static dichotomy has an adverse effect on the understanding of allegory itself. We cannot do away with the distinction, of course, and in the second part of this chapter allegorical negativity will be addressed, as well as the moves by which negativity comes to extend to the very limits of allegory, indeed, even into allegory's beyond. The hegemony of poststructuralist polemics in art theory tended to align the opposition of dialectics and deconstruction with, respectively, the modernist symbol and postmodern allegory. Accordingly, the symbol (in its more advanced forms) substitutes for dialectical mediation and sublation, whereas allegory calls up deconstructive discontinuity and deferral.[100] The conception of allegory itself also becomes situated between deconstruction and dialectics. It may not be accidental that de Man and Benjamin respectively occupy an ambivalent relation to these bodies of thought, and that neither are especially suited to proving dialectical thinking redundant. Although de Man distances his work from dialectics, he nevertheless takes its language seriously and often poses questions through a dialectical terminology. Indeed, contrary to those interpreters who see de Man's work as departing from an earlier dialectical-existential mode in the fifties and sixties to a more thoroughgoing deconstruction, some of his most dialectical moments can be found in his late writings. In fact, we might frame de Man's position in dialectical terms—a dangerous procedure, I realize, but not an unreasonable one: his emphasis is neither on the immediacy claimed within the established rhetoric of the symbol nor on sublation's mediated immediacy, but on the refusal of dialectical reconciliation or closure. We have already encountered the long tradition of such wariness toward closure and its frequent translation into an emphasis on the dynamic of negation. In fact, de Man—like Benjamin—proves to be sensitive to another dialectical feature: the dynamic and transferential character of allegorical thought. As we have seen, while the advocates of the symbol dismiss allegory's mechanical conception of the relation of sign and meaning, de Man, in contrast, argues that allegory entails the relation between signs.[101] The emphasis is placed not on a move into a world of signs but on the "relation between," and, as we will see, de Man explores the movements established by this relationality and stresses the epistemological consequences. At its best, dialectical thought approaches the mobility of thinking required to grasp these transpositions. It is a mistake, then, to found our understanding simply on the conventional characterization of allegory's disjunctive qualities. The familiar chains of attributes listed for allegory and symbol—and which we could continue to extend—have become too neat and static, and increasingly unhelpful. Disjunction is, after all, not only privileged as a quality of allegory; it also describes the conceptualization of its relation to the symbol. Significantly, despite providing the most articulate accounts of allegorical disjunction in their practice of writing and thinking, neither Benjamin nor de Man allow their arguments to conclude on this note. It is the sense of disjunction propounded in their writing that needs careful exploration and which has been largely overlooked in discussions of allegory and symbol. Another way of putting this is to ask how we can go beyond an ossified polemics to deal with the dynamics of the relations within these oppositions, their tensions and transpositions.

The Dark Background

In writings spanning his adult life, Benjamin identifies allegory with Baroque drama (the *Trauerspiele*), montage (in its Surrealist and its Berlin post-Dada modes), and the poetry of Baudelaire (its spleen). Common to all is distancing, discontinuity, and the use of rupturing devices. Allegory has a "destructive tendency," Benjamin argues in his late essay "Central Park."[102] Identifying it with "androgyny, the lesbian, the unfruitful woman," he sees allegory's intention as being to tear things from their normal situations and "to destroy the organic," and he appreciates the usability and adaptability provided by its inorganicism.[103] The politics that he came to attach to inorganicism can also be glimpsed in his late theses on history where the smooth historicism and progressivism of the social-democratic outlook is contrasted with the revolutionary "tiger's leap" and refusal of the fatalistic acceptance that this is how things are and how they must be.[104] The symbol, Benjamin argues in the *Trauerspiel* script, "remains persistently the same," whereas allegory is an idea, different from itself.[105] Each successive form taken by allegory becomes redundant; it is a contingent mode that foregrounds historical specificity, orientates itself to usefulness in the present, and registers discontinuity in the historical continuum.[106]

In the *Trauerspiel*, Benjamin focuses his analysis of allegory on seventeenth-century German mourning plays, such as the works of Andreas Gryphius, Johann Christian Hallmann, August Adolf von Haugwitz, and Daniel Caspers von Lohenstein (although he considers the dramas of the Spanish playwright Calderón to be the Trauerspiel's most "perfect form").[107] As we have seen, Benjamin is particularly interested in their "disjunctive, atomizing principle" and their use of *disjecta membra*.[108] For Benjamin, these plays follow a double or duplicitous logic, which can be identified at various levels: in the organization of the plays into action and interludes; in their double titles, which contain both literal and proper references; in their use of the word and the figural and in the overall tectonic structure of the drama.[109] He also finds the "duality of meaning and reality" echoed by the construction of the stage sets and the architecture of the theatres, which demonstrate disjunction in both spatial organization and plastic form, from scenography to a secreting of character in the "thousand folds of allegorical drapery."[110] Allegory is characterized by its accumulation of atomized linguistic, sound, or image ruins as dislocated signs, brought together with an "awkward heavy-handedness" such as the staccato-like commentaries that abandon any pretence to dialogue.[111] However, this emphasis on radical disjunction—not to mention the language that Benjamin employs (antithesis, division, duality, dichotomy, duplicity)—has to be considered more carefully. Whether he defines allegory externally (in relation to the symbol), or whether he discusses allegory's internal economy of negativity (the disjunction of sign and meaning), the oppositions he puts into play are never straightforward; instead, they need to be grasped by their dynamics. This holds even when—especially when—Benjamin explicitly refers to allegory's static character, its "rigidity" or "rigour mortis." (As he observes in "Central Park," allegory "offers the image of petrified unrest.")[112] Indeed, Benjamin suggests that *both* symbol and allegory became debased concepts in the hands of the symbol's advocates. The later Romantics and the Neo-Kantians, he argues, used this false symbol to evade art's ethical dimension. Although the concept of the symbol was made to sound dialectical by such writers, Benjamin insists that their "distorted conception of the symbol" lacked "dialectical rigour." Similarly, their dialectic of appearance and essence was, in fact, nothing but a paradox that failed "to do justice to content in formal analysis and to form in the aesthetics of content."[113] For the dialectic of form, Benjamin names the antinomies between convention and expression, secret and public expressions of authority, and cold technique and eruptive expression of interpretation. For the dialectic of content, Benjamin discusses the contradictory tendencies in the treatment of the profane detail in allegory, where it is both devalued (anything can mean absolutely anything else, thus making the detail unimportant) and elevated (the moment of the detail's sanctification on the higher plane).[114] We must beware, Benjamin warns, of treating the terms employed to describe allegory and symbol too rigidly.[115] Allegory's opposition to the symbol cannot simply be rendered as the opposition of thing to the personal, or of the fragment to the total; even Joseph von Görres's concept of the plastic, or real, symbol, which he opposed to the mystic symbol, comes closer, Benjamin insists, to something we might understand as allegory.[116]

The criticism of the symbolic mode is directed at the thought of Stefan George and Ludwig Klages, both of whom had considerable impact on the younger Benjamin. While remaining influential on his thought, they had become the object of his distrust after the First World War, exemplifying the approach that de Man calls "aesthetic ideology." According to Klages, every meaning should be immediately given as sensorial experience; the picture should be "unspeakable," attainable only by "insight" (*Schauung*), the access to which might be achieved through ecstasy, inebriation, or through the use of mind-altering drugs.[117] For Neo-Kantian thinkers such as Hermann Cohen and Carl Horst, in the symbol, as in the Christian creed, Being and sign were considered one; meaning was taken to be immanent to the symbol-object, requiring no mediation by representation or conceptual analysis. This immediacy was deemed to lie in the symbol-object's material presence and required no subject to activate its significance. In the *Trauerspiel*, Benjamin challenges the attempts of the Neo-Kantians to maintain allegory's subservient status and their failure to grasp "the dialectical quality of this form of expression":[118]

> The undialectical neo-Kantian mode of thought is not able to grasp the synthesis which is reached in allegorical writing as a result of the conflict between theological and artistic intentions, a synthesis not so much in the sense of a peace as a *treuga dei* between the conflicting opinions.[119]

These writers, Benjamin continues, took allegory to be "a harsh disturbance of the peace and a disruption of law and order in the arts."[120] (Gillian Rose also argues that the Neo-Kantians developed an antidialectical account of antinomy, a feature which she parallels with the approach taken by poststructuralist thinkers in the latter part of the twentieth century.[121]) Misunderstanding allegory's distinctive dialectical form, Neo-Kantian aesthetics saw allegory's ambiguity as something to be mistrusted.[122] Benjamin, in contrast, upholds this very quality. In his unpublished 1935 essay "Paris, the Capital of the Nineteenth Century," he also explains how ambiguity acquires a uniquely capitalistic form. Focusing on the figure of the prostitute, who is at once the commodity and its seller, he argues that such ambiguity should be understood as "the appearance of dialectic in images, the law of the dialectic at a standstill."[123]

In his "Epistemo-Critical Prologue" to the *Trauerspiel*, Benjamin criticizes the "attempt to find a substitute for reflection on the philosophy of art in a syncretism of cultural-historical, literary-historical, and biographical approaches." This syncretism tries to empathize with its object—creating a "pathological suggestibility" and engaging in "self-absorbed fantasizing"—where "idle curiosity masquerades as method."[124] Benjamin expresses his antipathy to those approaches that tried to flatten differences and silence opposing views. However, he warns, the Baroque totality is liable to propel its interpreters into dizziness. Therefore, in order to maintain critical distance it is important to take the mind "through a more or less ascetic apprenticeship, to the position of strength from which it is possible to take in the whole panorama and yet remain in control of oneself."[125] Benjamin castigates philosophy's dismissal of the detail. The love of beauty, he argues, is "empty dreaming" without "the life of the detail," and, moreover, "structure and detail are always historically charged."[126] Philosophical criticism should make "historical content" and "material content" the basis of its "truth content"; rather than holding up truth against history and material, the issue for Benjamin turns on illuminating the transformations of the one into the other. In short, allegory functions for Benjamin as a defense of dialectics, albeit a dialectic of a specific sort. (Whether this makes his project Hegelian is, I think, an open and complex question, depending on one's sense of Hegel, one's emphasis on method or result, and on orthodox or dissident reading, and one's take on the open dialectic versus a totalizing one.[127]) Benjamin's argument in the *Trauerspiel* is explicit, and a good deal has to be silenced to render it an attack on the dialectic. As we have seen, the Neo-Kantians cannot think dialectically, offering a symbol that is nothing but a pseudo-dialectic. The "baroque apotheosis," Benjamin maintains, "is a dialectical one"; the temporality of allegory is dialectical; a dialectical sense of allegory's "extreme form" is required to grasp the *Trauerspiele*; we require a dialectical understanding of allegory's antinomies, and a consideration of its dialectic of form and its dialectic of content.[128] Over a decade later in "Central Park," notable for its play on the qualities of the Aufhebung, he argues, "that which the allegorical intention has fixed upon is sundered from the customary contexts of life: it is at once shattered and preserved."[129]

The handling of antimonies in Benjamin's text is interesting. The dialectic of baroque allegory, he suggests, "is accomplished in the movement between extremes," variously described as that between exterior and interior, classicist structure and the means of expression, mathematical precision and uncontrolled imagination.[130] In the Baroque drama, Benjamin argues, the acts are divided by interludes or choruses that constitute allegorical assaults on the drama's "claim to be a Greek temple," emphasizing the "genuinely visual" (quoting Novalis).[131] The contrast between the diegetic flow of the acts (themselves constructed as a montage of "terraces") and the "expressive statuary" of the interlude's *tableau vivant* echoes, for Benjamin, a division of dream (or meaning) and reality.[132] Often seen by the critics as a deadening device causing an unnecessary drag on the action, these interludes, he suggests, are best understood not as immobility but rather as "the irregular rhythm of the constant pause, the sudden change of direction, and consolidation into new rigidity."[133] This sense of frozen or interrupted movement can be found throughout Benjamin's accounts of allegory: "dialectics at a standstill," "the image of petrified unrest," or even in his comparison of allegory with the stations on the Via Dolorosa.[134] The *Trauerspiele*, then, take on qualities akin to the children's games of "musical statues" or "peep-behind-the-curtain": there is a great deal of movement going on before the "freeze moment," and, it could be argued, just as much during it. The point to emphasize again, then, is not merely the sheer fact of antinomy, but the interchanges it sets in motion, and to raise our awareness of the quivering contingency and instability of the reified forms bequeathed to us by history.

Against the false faith in the symbol—whether Neo-Kantian or Neo-Classical—Benjamin presents allegory as an irrepressible negative power. Allegory is not simply a mode of representation that is an alternative to the symbol; rather, the symbol is overtaken by allegory: "Where man is drawn towards the symbol, allegory emerges from the depths of being to intercept the intention, and to triumph over it."[135] A good example of this transformation appears in his passage on J. J. Winckelmann's discussion of the Belvedere torso (see figure 3.1). Benjamin argues that in the very attempt to see and to describe the sculpture symbolically, Winckelmann's classicism must present it in allegorical terms: by exploring the torso part-by-part, Winckelmann's attention dissolves "the false appearance of totality."[136] Furthermore, in attempting to bring the torso into the realm of knowledge—a move that Benjamin associates with the Fall—Winckelmann necessarily evaporates the symbolic claim to transcend the concept. A few pages later, Benjamin describes how the "exaggerated realism" in the foreground of a

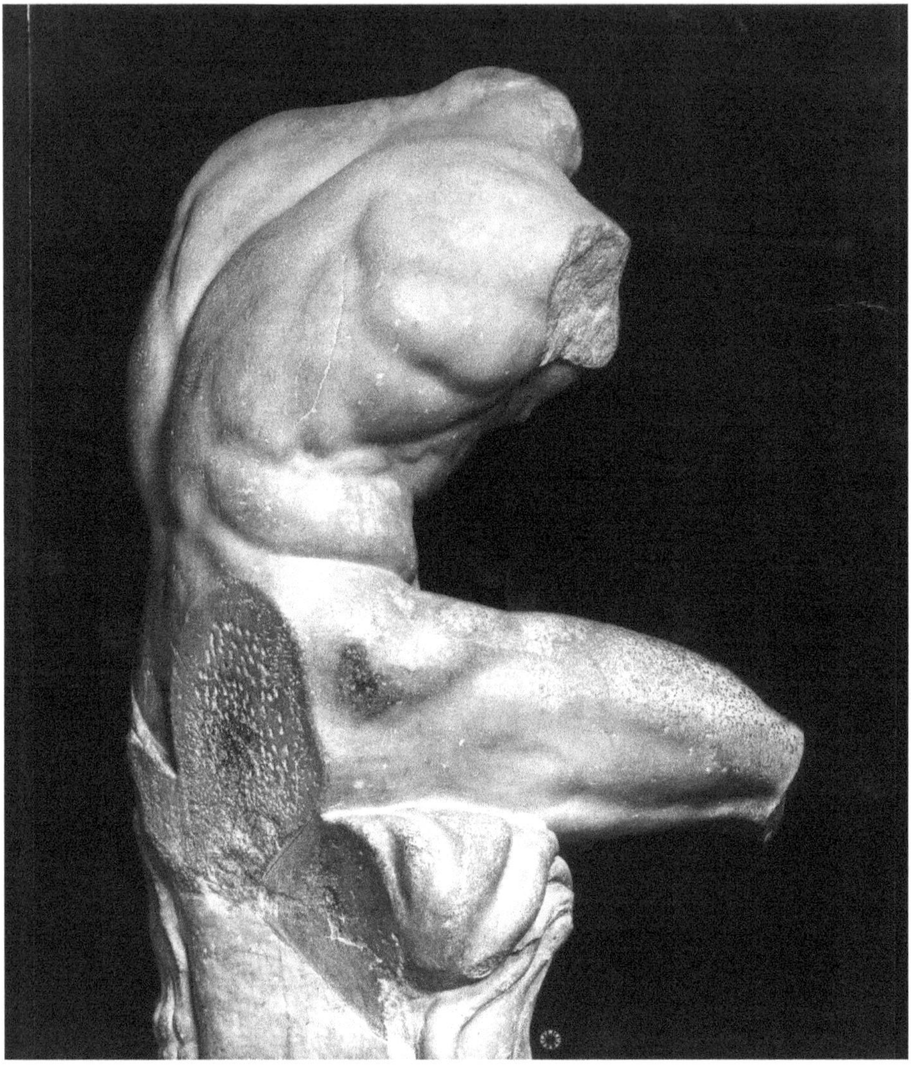

3.1 Belvedere Torso, Greco-Roman, marble, 159 cm high, Vatican Museums and Galleries, Vatican City, Italy. Photo: Alinari/The Bridgeman Art Library.

painting sets off the background's "visionary objects."[137] When Christ is depicted in the foreground, he can mediate Heaven and earth, but by being placed "in the realm of the provisional, the everyday, the unreliable" this same pictorial strategy also becomes a "provocative" and "offensive" gesture.[138] A significant feature of the Baroque, for Benjamin, lies in this desacralizing tendency, which runs counter to the period's explicit displays of religiosity (a trend that is given a specific twist in the post-Reformation German lands where the *Trauerspiele* represented the secularization of the mystery play).[139] What sets out in one direction finds its terminus at the end of quite another path, and repeatedly, we find that "the symbolic becomes distorted into the allegorical."[140] We can see another example of this when Benjamin addresses the question of writing and speech, a theme that has provided the object of so much attention in recent criticism. Benjamin commences with a polarity: "The division between signifying written language and intoxicating spoken language opens up a gulf in the solid massif of verbal meaning."[141] However, he treats this (now familiar) contrast in a distinctive way, and again his focus falls on the field of forces within the antinomy's extremes. "The antithesis of sound and meaning," he argues, "could not but be at its most intense where both could be successfully combined into *one*, without their actually cohering in the sense of forming an organic linguistic structure."[142] Baroque language, Benjamin claims, is "constantly convulsed by rebellion" of its elements; stripped of their original signification, "emancipated from any context of traditional meaning," the linguistic fragments of word, syllable, and sound nonetheless contain a threatening "remnant of meaning" and even diminishing echoes can serve as warnings.[143]

In considering the stage sets, he also describes how the dualities evident in the *Trauerspiele* acquire plastic form; the drop scenes allow a waver between foreground and background.[144] Benjamin notes how, in Baroque architecture, there is a division between the controlled, mathematical simplicity of the exterior and the exuberance of the interior's "uncontrolled imagination," which is not dissimilar to the antinomy he finds in the "cold, facile technique" and the "eruptive expression" of the *Trauerspiele* themselves.[145] The soaring angels of Baroque architectural sculpture, Benjamin tells us, are carried by massive pedestals and columns, which draw "attention to the difficulties" of support "from below."[146] To say this, of course, is like "foregrounding the device," but its effect goes beyond highlighting the artifice of representation. Benjamin's claims exceed his local subject matter: the baroque forces us to look "into the depths of language" and is particularly attuned to "the problematic character of art."[147] As his example makes clear, these architectural supports highlight "the difficulties" which *ground* the act of representation. Benjamin challenges the traditional relation between symbol and allegory where allegory appears as "the dark background against which the bright world of the symbol may stand out."[148] Ceasing to be the mere backdrop to the symbol's bright performance, allegory instead is revealed to be the symbol's *necessary* foundation. (De Man too describes allegory as more authentic than the symbol.[149]) What we see here is not the lightening flash of the symbol—the mystical instant, or the momentary insight, described by Creuzer; rather, it is allegory that sheds light onto the symbol's obfuscation, the flash of allegory that returns us to the play of history, nature, and ethics.[150] What is significant about this inversion is that it is not *just* a reversal of the standard valuation of the symbol-allegory opposition. Symbol and allegory are not treated simply as terms around which aesthetic preferences are staked. Benjamin's central claim is stronger: allegory (or more precisely the point to which allegory takes us) is the ground upon which we choose; and in choosing we either acknowledge or deny that ground.

Charity *Devoid of Charity*

Although there are evident differences between the two bodies of work, de Man's argument reveals similar patterns of argument to those we have encountered in Benjamin's writing. De Man addresses allegory and symbol most famously and explicitly in "The Rhetoric of Temporality" from 1969. The subject, however, frames most of de Man's writing, from his early essays through to his late work on the philosophy of Hegel and Kant. "Reading (Proust)," from 1972, is often taken as presenting de Man at his most clearly deconstructive, and as marking a further distance from his earlier more existential approach. His argument in the Proust essay turns on a passage concerning visual representation, and, as we will explore, he returns to this same episode in one of his very last essays. De Man takes as his object Marcel Proust's *À la recherche du temps perdu*, focusing on the account of Giotto's *Caritas*.[151] Celebrated for the richness of its metaphors and its imagistic language, Proust's book is traditionally treated as a prime example of the symbolic aesthetic. Through a close reading of the rhetorical structure of some key passages from the section *Du côté de chez Swann*, de Man subjects this assumption to a searching critique, insisting that the work neither offers "the unmediated experience of an identity" so celebrated by commentators, nor describes the rush of *mémoire involontaire*.[152] He argues that what is frequently seen as the reign of metaphor in this book is, in fact, shaped by the trope of metonymy. Thus, what seems to be symbolic is shown to be allegorical—a move that has a remarkable similarity to Benjamin's account of Winckelmann. Proust's text employs a wealth of "seductive metaphors," and even makes explicit statements with regard to the superiority of metaphor. However, de Man explains, "persuasion is achieved by a figural play in which contingent figures of chance masquerade deceptively as figures of necessity."[153] While a thematic or literal reading upholds metaphor as the defining trope in Proust, closer rhetorical attention to the textual performance reveals the limits such claims. (Fredric Jameson takes this to be de Man's most dialectical moment.[154]) According to de Man there are two incompatible readings of Proust: one based on the reader's aesthetic response, the other structured by rhetorical awareness. This contrast forms an aporia, which, as de Man insists, is based on a logical rather than a representational incompatibility. Commentators frequently praise the grand syntheses presented in Proust's work, in which the separate fragments are brought into the orbit of a master metaphor gravitating around "all the forces of" Marcel's life. However, for de Man this is merely the semblance of synthesis. According to him, the chain of metaphoric connotation (orchestrated around figures of internality, repose, darkness, and coolness) and the metonymic chain (externality, action, light, and heat) fail to add up.

Most of de Man's essay is devoted to elaborating this classic thesis of deconstructive criticism. However, it is the subsequent and final section of his essay where there is a change of gear and where the basic deconstructive lesson is subject to yet further interrogation. This additional move, largely unrecognized or ignored by de Man's readers, is very typical of his approach. De Man is not the first to note the role metonymy played in unpicking Proust's metaphors, and he acknowledges the significance of the accounts offered by Gérard Genette and Gilles Deleuze, which emphasize metonymy, allegory, and disjunction. De Man considers their proposal—one which might be compared to that of Owens—that we read *La recherche* as an allegory of its own deconstruction.[155] It is this that the latter part of de Man's essay aims to challenge. For de Man, to say that Proust's novel is an allegory of its own deconstruction is to reinstate coherence. In other words, this kind of deconstructive reading may undo and disrupt metaphoric or symbolic structures, yet, however much it upsets a secure epistemology with a negative one, it simply stabilizes the reading at a higher level. As de Man argues: "What is at stake is the possibility of including the contradictions of reading in a narrative that would be able to contain them . . . an allegory of reading."[156] By treating allegory as "the report" of the process, de Man argues, its "destructive power" is evaded; the "contradictions of reading" are contained, grounding "the stability of the text."[157] As he puts it: "at the far end of its successive negations," the interpretation "will recover the adequation between structure and statement on which any thematic reading depends."[158] It is precisely this secure deconstruction that de Man proceeds to undermine. Elsewhere he distinguishes "primary deconstructive narratives centred on figures and ultimately always on metaphor" from allegories that work at "the second (or the third) degree." For de Man, irony and allegory are not playful modes. To unpack and deconstruct a symbolically saturated text is not a consoling act or triumphal intellectual achievement; rather, this activity always points to something deadly serious. Although he shares many of the same targets, he challenges those deconstructive accounts that claim positions of superiority over their objects. For de Man, not only must symbolic and mimetic claims be problematized, so must allegorical ones. De Man's thought shares with thinkers such as Adorno a concern with epistemological exploration and a desire to push negative moments further. In this sense, certain dialectical accounts may come closer to de Man's version of deconstruction than the work of many self-professed deconstructionists. Christopher Norris explicitly parallels de Man's approach with that of Adorno: "Deconstruction is indeed a form of negative dialectics, an activity that carries on the project of immanent or self-reflective critique developed by Hegel out of Kant, but which turns this project against its own desire for such premature endpoints as Symbol or Absolute Reason."[159]

Indeed, de Man's approach to allegorical negativity entails allowing a degenerative potential to take hold. This is where "Giotto's *Charity*" becomes significant for his argument. In this episode from *Swann's Way*, Proust's fictional narrator, Marcel, considers photographs of Giotto's famous cycle from the Arena Chapel (see figure 3.2). As de Man reminds us: "Proust does not start out from a direct encounter with Giotto's frescoes, but from John Ruskin's commentary on Giotto's Vices and Virtues of Padua."[160] In a section of *Fors Clavigera*, Ruskin notes how he had initially misinterpreted Charity's stance: he first saw her as receiving God's heart, but he subsequently came to see God as the re-

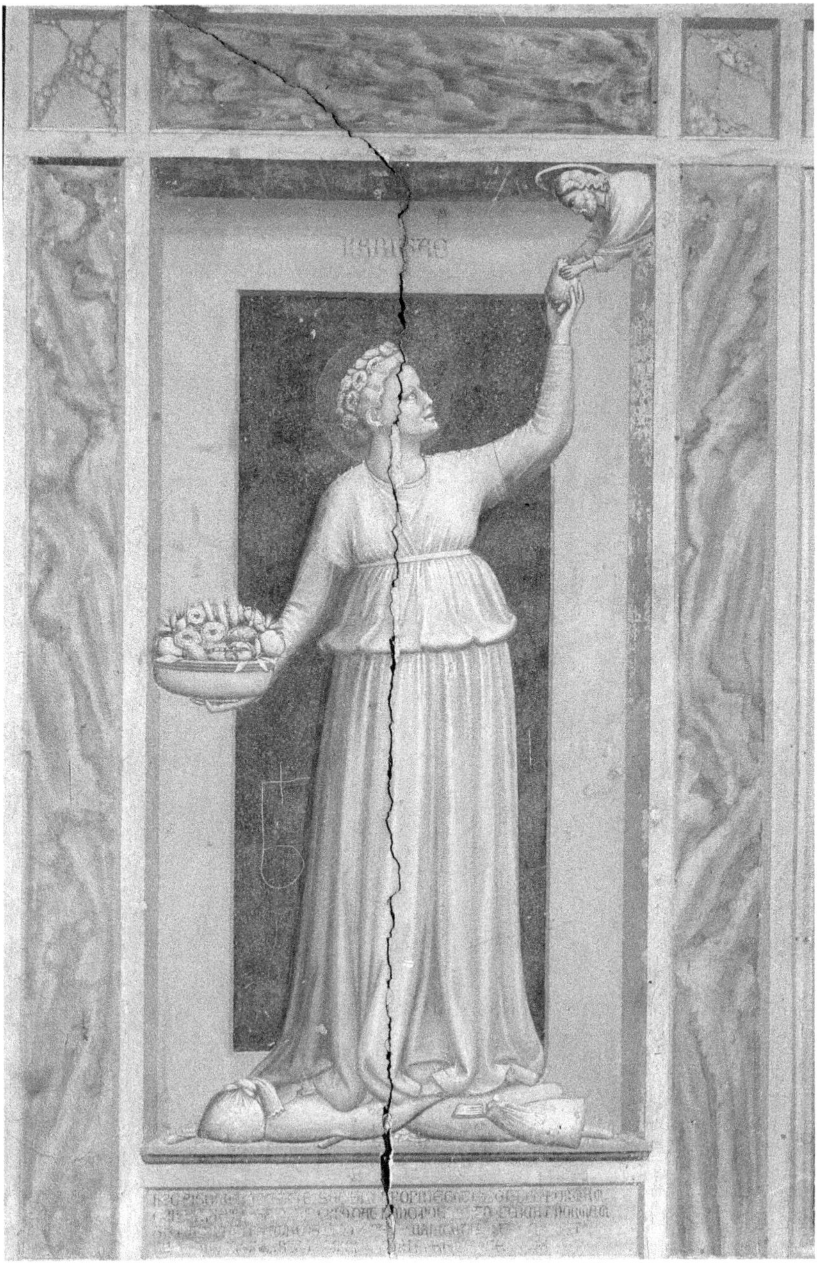

3.2 Giotto di Bondone (c. 1266–1337), *Charity*, c. 1305, fresco, Scrovegni (Arena) Chapel, Padua, Italy. Photo: The Bridgeman Art Library.

cipient of her heart.[161] Marcel's reflections are prompted by thoughts about the pregnant kitchen maid who works at his family home in Combray. Swann, a family friend (and who had been the bearer of these photographic gifts) had nicknamed the kitchen maid "Giotto's Charity." Marcel contemplates several ways of seeing their resemblance and nonresemblance. In Swann's hands, "Giotto's Charity" is simply a metaphor for the kitchen maid; he sees in the maid a physiognomic resemblance to the woman depicted in the fresco. For Marcel the similarity between maid and fresco appears altogether more complex: it turns on the inability of each character—the maid and the painted Virtue—to grasp her own significance and, more particularly, on the disjunction between vehicle and its proper, or allegorical, meaning. In the fresco Charity carries a vessel and Marcel thinks that the pregnant kitchen maid is blind to the spiritual import of her own "mysterious basket." At this point, his interpretation of the real and allegorical cross over. Marcel notes that the maid is just one of a succession of domestic workers employed at Combray. In this sense, a specific kitchen maid comes to signify abstractly and figurally as a role or type. Meanwhile, Marcel believes that the power of Giotto's allegorical fresco resides in its realistic depiction. An abstract and sacred personification resolves itself into an everyday scene of secular particularity. The kitchen maid and the fresco of Charity are representations, de Man observes, both of which require reading.

In de Man's account, Proust takes the reader from symbol to allegory by switching from Swann's perspective to that of Marcel. We might break down this transition; here it is necessary to engage in detailed reading of the passages in de Man. As we have already noted, the relation of literal and figural meaning is played out across the figure of the kitchen maid and her resemblance to the fresco. Although he lacks the terms with which to describe it, Swann posits this relation as metaphor and as synthesis in which oppositions—particular and universal, matron and virgin, profane and sacred, low and high—are reconciled. Marcel, who is described by de Man as more rhetorically sophisticated than Swann, sees not symbolically but allegorically. He does not identify a *synthesis* of literal and figural elements, but rather a *disjunction*: a divergence between a meaning proper to the literal and a meaning proper to the allegory. This discussion focuses upon the fresco images, with only passing reference to the kitchen maid (and increasingly to a comparison with the cook Françoise).

Initially, however, Marcel does not like this discordance between their look and their meaning, which seems so pronounced, so that Charity was "Charity devoid of charity," *Justice* looked more like Injustice, and *Envy* (*Invidia*) failed to invoke the vice and just reminded Marcel of a medical illustration (see figure 3.3). These discordances hinder Marcel's attainment of the sort of symbolic reconciliation that Swann, with his limited rhetorical skills, is able to reach. Consequently, Marcel at first deems the frescoes awkward, making them into failed symbols, and thereby allegorical, in their inability to synthesize. At this point, however, he realizes that the discordance ought not to be seen as a shortcoming but as the core of the frescoes' power:[162]

> I came to understand that the arresting strangeness, the special beauty of these frescoes derived from the great part played in them by symbolism, and the fact that this was represented not as a symbol (for the thought symbolised was nowhere expressed) but as a reality, actually felt or materially handled, added something more precise and more literal to the meaning of the work, something more concrete and more striking to the lesson imparted.[163]

It is this power—which now resides precisely *in* disjunction—to which Swann, in his ignorance of the difficulties, is oblivious.

It is crucial to grasp that Marcel's reading is not just a *revaluation* of allegory; it does not place a plus where once a minus had been (and here we see the difference of this reading from a postmodern "allegorical impulse" that inverts the rule of the symbolic). According to de Man, Marcel's shift in appreciation is quite distinctive. It rests neither on seeing the frescoes as symbols, nor simply on perceiving the disjunction of "Charity devoid of charity." Instead, it depends on Marcel's recognition that, no matter how divorced their relationship might appear, the vehicle (the literal representation) is necessary for the conveyance of meaning, a necessity working across the disjunction. The fresco's meaning is made more forcefully because of the realistic representation that seems "devoid of charity." Allegory, asserts de Man, cannot do without the powers of literal representation; the allegory is dependent upon a literal representation that is disjunct from the proper meaning. Marcel goes on to note that in real life the "truly saintly embodiments of practical charity" never appear remotely compassionate, but behave in a brusque and matter-of-fact way, as if they were each "a busy surgeon."[164] In this discussion, there

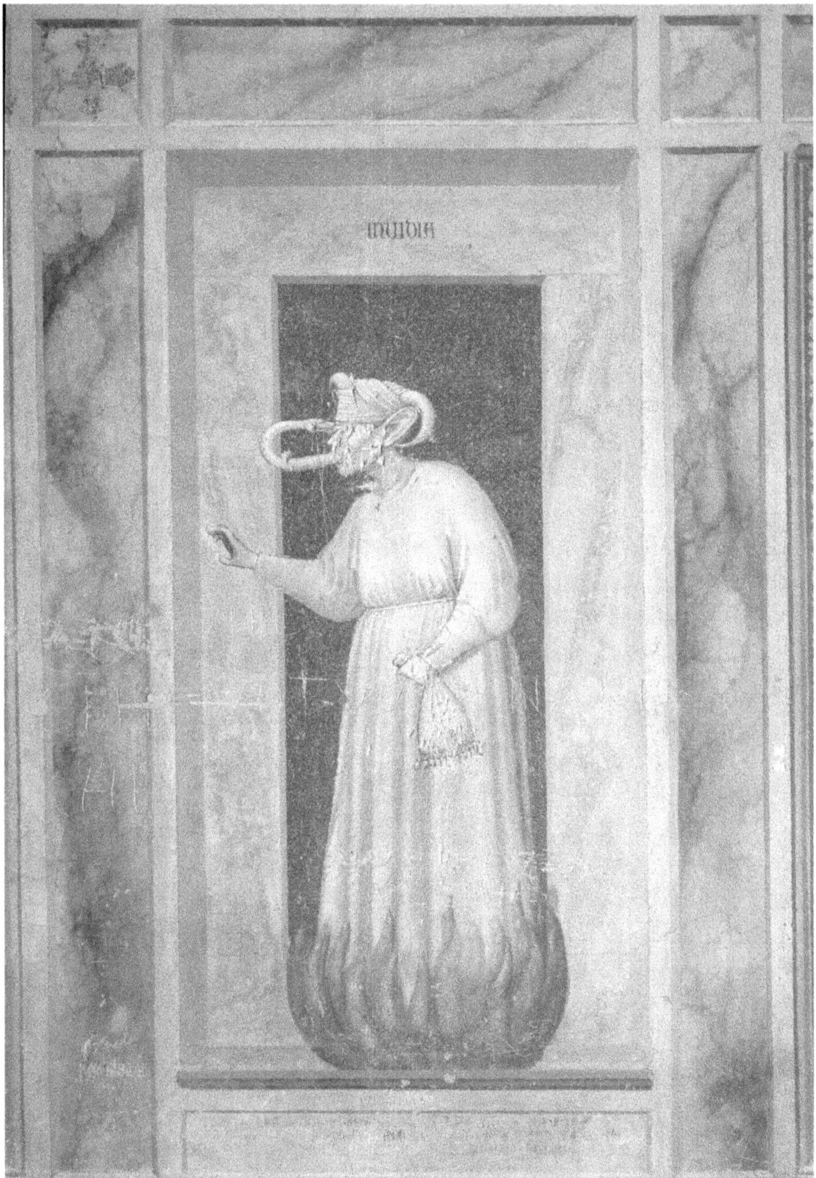

3.3 Giotto di Bondone (c. 1266–1337), *Envy*, c. 1305, fresco, Scrovegni (Arena) Chapel, Padua, Italy. Photo: The Bridgeman Art Library.

is no simple opposition between the literal and allegorical meanings, but something like a dialectic of mutual dependence.

In this final reading of allegory, the emphasis is placed on an open dialectic of connections and transformations. Allegory is not just the fact of disjunction but a process in which tensions are exacerbated, where antinomies cross, interact, and even deteriorate. However, the interaction of necessity and disjunction at work is not simply one of mutual support, so much as one of *destructive* dependence in which a corrosive dynamic is set in motion. In his account of *Charity*, de Man describes a deflection through which the literal representation overtakes the proper allegorical meaning. With Giotto's *Invidia*, an iconic detail is hyperbolized: the exaggerated emphasis on the serpent tongue almost obliterates the proper meaning—and we might extend the example to Envy's ears. We look at Envy, but we do not meditate on the evil implications of envious tendencies (the proper meaning), nor, in response, do we turn to goodness and virtue (the proper intent of the allegory).[165] De Man puts it this way: "the allegorical representation leads towards a meaning that diverges from the initial meaning to the point of foreclosing its manifestation."[166] The force of fixation upon such details, he argues, redirects our thoughts: "the mind is distracted towards something even more threatening than vice, namely death."[167]

Negative Capability of Infinite Regressions?

The same destructive dynamic parallels the sense of irony described by de Man in "The Rhetoric of Temporality."[168] Here he notes how irony "reveals the existence of a temporality that is definitely not organic, in that it relates to its source only in terms of distance and difference and allows for no end, for no totality."[169] He claims that both allegory and irony "are determined by an authentic experience of temporality which, seen from the point of view of the self engaged in the world, is a negative one."[170] This account of irony repeats themes already encountered in de Man's discussion of allegory: inorganicism, distance, difference, irreconcilability—in short, some form of radical negativity. De Man draws out the distinction between Baudelaire's conception of irony (*le comique absolu*) and "simple comedy" (*le comique significatif*). The latter operates at an intersubjective level; it is orientated toward others for its effect and entails laughing *at* someone. In *comique absolu* this relation is turned inwards, becoming a matter of self-duplication or *dédoublement*, where the self assumes a consciousness capable of laughing at itself. What is treated as the interaction of two different subjects in simple comedy becomes, in *comique absolu*, that between two parts of the same subject—or between the two parts of a divided or split subject where one part reflects ironically on its other part. De Man implies that *comique absolu* is a better approximation of the groundlessness of negative self-knowledge, and that simple comedy entails a retreat from that knowledge. This high degree of self-reflexivity is seen to be the preserve of sophisticated, rather than everyday, language use (that is to say, it is the language use of philosophers and artists), a point which concurs with Benjamin's observation that Baroque allegory is a privileged and elite mode of language. De Man clarifies the full implications of this:

> The reflective disjunction not only occurs *by means of* language as a privileged category, but it transfers the self out of the empirical world into a world constituted out of, and in, language—a language that it finds in the world like one entity among others, but that remains unique in being the only entity by means of which it can differentiate itself from the world. Language thus conceived divides the subject into an empirical self, immersed in the world, and a self that becomes like a sign in its attempt at differentiation and self definition. . . . The ironic language splits the subject into an empirical self that exists in a state of inauthenticity and a self that exists only in the form of a language that asserts the knowledge of this inauthenticity. This does not, however, make it into an authentic language, for to know inauthenticity is not the same as to be authentic.[171]

In other words, irony's self-consciousness does not allow a recovery of the subject's coherence, and de Man is critical of Szondi and Starobinski for making the "irony of irony" into a recovery (thus avoiding its absolute and unnerving giddiness). According to de Man, the movement of ironic consciousness induces an "unrelieved *vertige*," which cannot be corralled except at the expense of reducing it back into *comique significatif*—that is, it cannot be contained without the loss of distance between fiction and the empirical world entailed by intersubjectivity.

Depending on how one reads the histories, allegory in the modern era either declined or was fundamentally transformed. Northrop Frye believes that the modern period was "antiallegorical," whereas Angus Fletcher argues that allegory proliferated but through antiaffirmative, negative, regressive, katagogic, "kako-daemonic," or evil-daemoned forms.[172] According to Honig, the process of self-discovery characteristic of premodern allegorical literature, through which the hero "extend[s] his identity" to regain paradise, is transformed by modern allegory into a journey with "no vital mission": the narrative "constantly dwindles," and "instead of finding his many actual identities," the hero "shrinks and is finally converted into nothingness."[173] Gay Clifford also points to how the affirmative functions of theocentric allegory give way in the modern period to anthropocentric and subversive uses.[174] Similarly, Quilligan refers to "the vertigo readers of allegory often suffer when the plot simply evaporates."[175] Modern allegory, she argues, is characterized by doubt, and by the lack of optimism that this doubt—linguistic and ethical—can be "overcome."[176] In this sense, modern allegory is marked not only by the wavering of spiritual doubts encountered on the journey; in the modern form, doubt, skepticism, and solipsism become all-engulfing. The seemingly rarefied rhetorical debate concerning symbol and allegory, then, takes shape within a wider concern with the crisis of legitimation.[177]

The abyss engendered by forgetting thought's groundlessness, like irony's "permanent parabasis," is a perpetual dialectic of self-destruction and self-invention with no reconciliation or closure.[178] Although, in Hegel's account in the *Phenomenology*, we move through progressive stages of self-consciousness, de Man suggests (and here he runs counter to many of Hegel's commentators), that this is a process without any resolution, consisting only of repetitions and increasing negative self-knowledge. The "progressive" dialectic, then, is rendered as "regressive" and entropic, and, according to de Man, the *Phenomenology* progresses on the basis of an implicit acknowledgement of this destructive dynamic. De Man notes how:

> Irony divides the flow of temporal experience into a past that is pure mystification and a future that remains harassed forever by a relapse into the inauthentic. It can know this inauthenticity but can never overcome it. It can only restate and repeat it on an increasingly conscious level, but it remains endlessly caught in the

> impossibility of making this knowledge applicable to the empirical world. It dissolves in the narrowing spiral of a linguistic sign that becomes more and more remote from its meaning, and it can find no escape from this spiral. . . .[179]

He suggests that, "even more clearly than allegory," irony takes us to a consciousness that is "an unhappy one."[180] The reference here is again to Hegel's concept of the unhappy consciousness, which has already been encountered in the discussion of Clark. In contrast to the externally driven dialectic of lordship and bondage, the unhappy consciousness internalizes negativity into split consciousness (for Hegel, a split between Subject and Substance).[181] As with the distinction between simple comedy and irony considered above, the moment of internalization exacerbates the effects of this division. The figure of the unhappy consciousness occurs elsewhere in de Man's work and typically his readings focus on preventing a condition of negation being turned into affirmation. He notes, for instance, how Serge Doubrovsky treats the unhappy consciousness as a condition of "plenitude," and therefore lacks the "dialectical anxiety" of his intellectual model, Maurice Merleau-Ponty.[182] The issue here is not the potential existential thematic, whether this is identified specifically with de Man's earlier work or with "anthropological" accounts of the master-slave dialectic developed by Kojève. What is important is the transformations registered in de Man's version of aporia, the constant struggle to prevent nonidentity from being resolved into identity at another level, to transcend irony and become metaironical, "without falling into the myth of an organic totality."[183]

In his diary entry for the January 18, 1926, inscribed while he was in Moscow with Asja Lacis, Benjamin notes:

> Then I read the lesbian scene from Proust. Asja grasped its savage nihilism: how Proust in a certain fashion ventures into the tidy private chamber within the petit bourgeois that bears the inscription *sadism* and then mercilessly smashes everything to pieces, so that nothing remains of the untarnished, clear-cut conception of wickedness, but instead within every fracture evil explicitly shows its true substance—"humanity," or even "kindness." And as I was explaining this to Asja, it became clear to me how closely this coincided with the thrust of my baroque book. Just as the previous

> evening, while reading alone in my room and coming across the extraordinary passage on Giotto's *Caritas*, it had become clear to me that Proust was here developing a conception that corresponds at every point to what I myself have tried to subsume under the concept of allegory.[184]

Like de Man, Benjamin describes modern forms of allegory as etiolating. He also emphasizes the process of internalization: "Baroque allegory sees the corpse only from the outside. Baudelaire sees it also from within."[185] Buci-Glucksmann observes how Benjamin's account of Baudelairean allegory incorporates death and abyss as spleen; that the corpse is now inside oneself, she notes, is "a radically disturbing novelty which demolishes the acquired certainties of the 'subject.'"[186] In fact, a similar withering process, here attributed specifically to modern allegory, can also be found in Benjamin's account of the *Trauerspiele*. Among the range of antinomies he associates with these plays, he analyzes the "elegant antitheses" of Baroque metaphor, notes the juxtaposition of sensuous metaphors with "an extreme recourse to concrete words," and alerts us to a passage of Cysarz in which abstract and concrete are put into play: "Every idea, however abstract, is compressed into an image, and this image, however concrete, is then stamped out in verbal form."[187] Traditional poetic metaphors behave contrary to expectation; they fail to deliver their meaning. Indeed, the "visuality" implied by such loaded metaphors, their imagistic use of language—we might say their *excess* of metaphoricity—serves to undermine metaphor itself. (Recall Clark's account of Pollock's art, which in its effort to sidestep metaphor, resorted to metaphor's multiplication.) The intensity of these metaphors makes Baroque language "heavy with material display" or "intent on the display of its own substance."[188] Instead of a transcendent image, Benjamin notes how "With every idea the moment of expression coincides with a veritable eruption of images, which gives rise to a chaotic mass of metaphors" (a figure perhaps adapted by Tafuri in his description of Piranesi's *Campo Marzio*, not as a late-Baroque "tumult in the whole" but as a "monstrous pullulation of symbols devoid of significance").[189] As their imagery becomes uncontrollable, the metaphors wither and "ideas evaporate in images."[190] A little later, Benjamin observes how the language of the *Trauerspiele* "expands in painterly fashion in the alexandrine."[191]

Allegorical Decision

It would seem at first sight that allegory is radically divorced from any commitment to the real. The poet, Quilligan claims, does not mimic life, but "the life of the mind" and attends to questions of language.[192] However, she argues that the text is not solely an intensive internal consideration of language; reading must nevertheless point to a world beyond itself and, in contrast to Foucault's much-cited account of Borges as an example of a self-referential text, Quilligan insists that allegory does not form a closed system.[193] Indeed, the point of allegory's attention to language is not to take us deeper into linguistic self-referentiality but to question its reification and our "culture's assumptions about the ability of language to state or reveal value."[194] She writes:

> By virtue of the fact that the immediate focus of the narrative is the language in which it is written, not only must the reader come to terms with the language in which such questions are asked, but he must also recognize that his answers—or such answers as seem to be indicated by the text—can be made only in language. The circular process ends in a self-consciousness the only way out of which may seem to be an arbitrary act of choice. Language does or does not lie. And if a reader chooses not to choose, he or she is left with a series of infinite regressions.[195]

Language has to be scrutinized, the better to understand its "slippery tensions between literalness and metaphor."[196] Presented with the "nearly fatal mistakes" made by the characters, the readers come to recognize their inability to distinguish "bad literal-mindedness" from "good literalness"; but the attentive reader is likewise confronted with her own tendency to misread.[197] Quilligan concludes that allegory's probing of the limits of language aims to raise the reader's awareness and to provoke "ethical action."[198] The "burden of choice" proffered by the allegorical text—true or untrue—extends, she argues, to the extratextual world where the choice becomes that between optimism and dwelling in "a gloom of negativity."[199] Quilligan concludes with Dante, who noted that those who refuse to decide in what they believe also reject the moment of self-definition; they "are condemned to spend eternity chasing elusive banners in the vestibule of the inferno. Not even hell will have them."[200] While this burden of choice clearly echoes earlier Christian forms of allegory—to opt for good over evil, to be moved to repentance and salvation—the parallels are limited; it is not so straightforward for "World" and "Language" to step into the shoes once worn by "Good" and "Evil." The question of choice and an ethical dimension to allegory is evidently present in Benjamin's account. Allegory takes us beyond the symbol, but in itself this is not sufficient. Allegory demystifies the symbol, he argues, and it better approximates the truth of a fallen world by recognizing the disarticulation of consciousness and *physis*. The symbol mistakenly takes truths as immediately present and plenitudinous; for allegory, they are acknowledged as removed and negative. Yet, despite this promise of demystification, Benjamin suggests that allegory too is limited: it fetishizes fragments and retreats to the inwardness in which the allegorist contemplates, and grieves over, a world devoid of meaning. Unhindered negativity—even at its most dynamic—leads only to "melancholic immersion," and Benjamin argues that just as "those who lose their footing turn somersaults in their fall, so would the allegorical intention fall from emblem to emblem down into the dizziness of its bottomless depths."[201]

The retreat into self-referentiality is, according to Hegel, characteristic of irony. Here the "I" sees itself as "lord and master," but, by its very inwardness, it becomes "inadequate to itself."[202] In an attempt to mitigate this inadequacy, the ironic "I" craves solidity and substance, yet remains unable to renounce the "inner harmony" that it derives from its ironic condition, remaining in self-imposed internality and inactivity.[203] Similar problems beset the "morbid beautiful soul."[204] They also come to haunt the unhappy consciousness, which "can only find as a present reality the grave of its life."[205] The relapse into self, he writes later, means "the return of consciousness into the depths of the night . . . a night which no longer distinguishes or knows anything outside of it."[206] For Hegel, this threatens a dissolution into the pure identity of "I = I," for which the reached-for antidote is yet more self-absorption, but which only risks further self-erasure. The exit from the "night" of pure identity is figured through the fall of the "Son of Light."[207] Lucifer's expulsion from Heaven functions as the necessary moment that prevents our complete subsumption in nature—the emergence of humans from the "archaic" as Benjamin or Adorno might say—and his fall represents the point where "immediate existence suddenly changes into Thought."[208] While "Evil in general is self-centred being-for-self," Hegel argues, "Goodness is what is simple and without self."[209] Without differentiation there can be no active (only sensuous) consciousness—and, therefore, no possibility of freedom. The rent condition of "Evil," then, is posited as a necessary price to be paid for consciousness and it is the necessary ground of freedom. (Not for nothing did Lukács alert us to Friedrich Schlegel's accusation that charged Hegel's dialectic and his "system of negation" with being "actual philosophical Satanism."[210]) Thus, there is an irony to the condition of both the ironic "I" and the model of "self" described as the unhappy consciousness. Both are split; both result from the primal separation. Yet in the face of the extremities of self-referentiality, at the point of a hyperemphasis of free self, a paradoxical situation arises: separation threatens to return into identity. (This is, essentially, the model followed by Adorno and Horkheimer in their account of modernity.)

Following this line of thought, we should reflect on de Man's argument. His insistence on the moment of nonidentity, if hypostatized, leads to its inverse: the dissolution into pure identity—and represents the real threat of self-erasure. The fundamental question of allegory turns on this double bind of a hypostatized nonidentity that—by way of its very insistence on separation—threatens to collapse back into an equally rigid identity. The fall of Lucifer inaugurates "three original satanic promises," Benjamin argues in the *Trauerspiel*, the illusions of freedom, independence, and infinity, which are thematized in allegorical works. He draws our attention to another aspect of this inflexible hypostasis:

> The absolute spirituality, which is what Satan means, destroys itself in its emancipation from what is sacred. Materiality—but here soulless materiality—becomes its home. The purely material and this absolutely spiritual are the poles of the satanic realm; and the consciousness is their illusory synthesis, in which the genuine synthesis, that of life, is imitated.[211]

In the final pages of his *Trauerspiel*, Benjamin charts allegory's "about-turn into salvation and redemption" where "the direction of allegorical reflection is reversed."[212] Thus, Benjamin's approval of allegory's "destructive" work needs to be situated.[213] In 1931, he wrote about the nihilism current among Left intellectuals:

> this left-wing radicalism is precisely the attitude to which there is no longer, in general, any corresponding political action. It is not

> to the left of this or that tendency, but simply to the left of what is in general possible. For from the beginning all it has in mind is to enjoy itself in a negativistic quiet.[214]

For Benjamin, it is clear that another response must be found and it has been widely acknowledged that he advocates a move into praxis. Edwin Honig also voices circumscription over allegory's plunge into the "abyss of despair." Allegory, he insists, does not entail "complete stultification" but provides the catalyst for a "leap" of commitment. In the end, allegory must abandon its "fundamental device of hypothetical construction" and opt for life, and "the allegorical work dispenses with the concept of allegory."[215] We are, then, confronted with a choice as to how to read allegory itself: on the one hand, it can be understood to point beyond its own melancholic form to the "world," the extratextual, construction, or praxis; on the other, it seems to offer infinite regress. The problem might be presented in different terms: as a resolution akin to the Hegelian Aufhebung, or as a form of perpetual openness. If the former is often taken as some form of return to order (the models being the Prussian state which was claimed as the apotheosis of Hegelian philosophy, the historicism of social-democracy, or the orthodoxies of Stalinist Diamat), it can also be figured as constructive, as a progression to building a social framework beyond capital. Similarly, while open negativity can be an emancipatory impulse, it can also signify the drift either into torpor, fatalism, complacency, nihilistic quiet, that is, as emancipation's hobbling.

Signs, Symbols, and Some Strange Turnings

At first sight, De Man's writing may seem less easy to accommodate into the account of allegory just outlined, but, as we have seen, he warns against resolving allegory into an allegory of its own deconstruction. Already in "The Rhetoric of Temporality," he notes that while allegory and irony are disenchanted forms of language, and while they are "de-mystified when they remain within the realm of their respective languages," they become, however, "totally vulnerable to renewed blindness as soon as they leave it for the empirical world."[216] Norris insists that de Man "allows of no premature appeal to rhetoric as a means of escape from epistemological issues."

> Rhetoric may figure as the problematic term, the aspect of language that complicates the move from phenomenal perception to concepts of pure understanding. But it can exert this deconstructive leverage only in so far as it remains an activity of thought closely in touch with epistemology and critical reason.[217]

Accordingly, Norris draws a firm line between de Man's approach and poststructuralism, and Andrzej Warminski notes that it is "far more radical and far more precise" than most deconstruction recognizes.[218] Jameson goes as far as to argue that the theories of de Man and Derrida "have nothing whatsoever to do with each other."[219] Whereas Derrida tries to imagine how to pose in language a prelinguistic moment or inaccessible radical difference, de Man addresses "the birth of abstraction and indeed philosophical conceptuality as such."[220] Thinking especially of de Man's writing on Rousseau, Jameson considers his work to be a struggle with "restoring the clumsiness of some initial thought processes" and he sees de Man's project as "returning to the act of thinking as praxis."[221] By recovering the awkwardness of the activity of thought, Jameson argues that de Man sought to strip "away the reifications that sediment around that act when it has become an object."[222] Jameson locates metaphor as emerging from this moment of conceptual abstraction, and sees allegory's self-reflexive mode of narration as emerging from "the primal metaphorical dilemma."[223] Jameson recognizes that de Man's interest in rhetoric and tropes is not narrow, but rather focuses specifically on "the dynamics of the trope." Appreciating the attention to thought as praxis, Jameson objects to the way de Man's critique of metaphor is a "dethroning" (albeit one which does not enthrone another trope in its stead) and how he uses metaphor to enact "transcoding" between philosophy, politics, literature, and tropology.[224] Metaphor's tropic turn is treated here as mediatory, but this is a mediation in which distinct fields of theoretical discourse are identified and "pronounced to be 'the same' as" one another. For Jameson, this makes de Man's metaphor "itself a metaphorical act and a violent yoking together of distinct and heterogeneous objects."[225] The nature of this connection, it could be argued, lies at the heart of allegory. Honig notes that it is allegory's working with "identification, motion, unity, and immutability" that causes it to constantly appear "on the borders between religion or philosophy and art" and those "between primitive mythological figurations and the more sophisticated structures of philosophical thought."[226]

Just as Jameson concerns himself with the clumsiness of thought in de Man, T. J. Clark, as touched on in the first chapter, probes how the strange awkward artificiality of Cézanne's paintings of bathers interrupts aesthetic expectations.[227] Clark is here in dialogue with Fred Orton's work on Cézanne, both drawing on one of de Man's late essays, "Phenomenality and Materiality in Kant."[228] The discussion turns on the materiality of the artistic practice, alluding to the *German Ideology*: the "practical activity" of painting invokes Marx and Engel's discussion of language as "practical consciousness." Seeing de Man as the thinker not only of allegorical disjuncture but also of thought's "grating and locking" and as a theorist of material recalcitrance returns us to a position not too far from Benjamin's emphasis on praxis and his argument that allegory resurfaces from the depths of the symbol. Emphasizing that Cézanne's work is best seen as a practical engagement with "the powers and limits of a practice of knowledge," Clark presents the artist as stretching his symbolic ambitions to their breaking point. By way of his sustained and practical attention to what is possible with representation, Cézanne's intention to fulfill a symbol aesthetic—which in the ideological form was articulated by Joachim Gasquet—disaggregates into the allegorical.[229]

It is one thing to find in de Man attention to the praxis of thinking, even if the sensuousness of the activity is characterized less by a traditional notion of aesthetics and more by a halting, uncomfortable grappling; and it is one thing to allude to the materialisms of Marx, Bauer, or Feuerbach. But what about the Hegelian dialectic itself? In an early essay from 1955, de Man notes:

> If there ever was a philosophy of necessary separation, it is Hegel's; to assimilate the notion of Absolute Spirit with idealist reconciliation is to simplify all the way into misprision.[230]

In his late writings from the nineteen-eighties, he not only addresses Kant but also returns to explore this "necessary separation" in Hegel.[231] Considering the *Aesthetics*, de Man explores "the commanding metaphor . . . of interiorization" in Hegel's "dialectics of internalization," which he takes to be the most powerful presentation of symbolic rhetoric and the aesthetic. Here de Man identifies the concept of *Erinnerung*, or remembrance, at the heart of the theory of the symbol. As the "recollection as the inner gathering and preserving of experience," the emphasis of Erinnerung is on the organicism of the relationships it mediates and the interpenetration of form and content.[232]

However, while de Man presents Hegel as the ultimate theoretician of Erinnerung, he also argues that he is also the philosopher of *Gedächtnis*—that is, the type of memory, or memorization, associated with rote learning and with Proustian *mémoire volontaire.* Unlike the organicism of Erinnerung, the relation of name and meaning in Gedächtnis is mechanical and arbitrary; like allegory and the sign, it proposes an "empty link." De Man notes that it is Gedächtnis not Erinnerung—rote memorization rather than the fullness of recollection—which, despite the problems it introduces to his system, plays the key role in Hegel's thought. Needless to say, this is not the Hegel we are accustomed to encountering. De Man's argument is involved and complex: the discussions of the aporias of deictic language, and of the necessary self-effacing of first-person pronoun are particularly fascinating, taking us to passages in the *Encyclopaedia Logic* and, by implication, back into the discussion of Sense-Certainty in the *Phenomenology.*[233] In his argument, we find some of de Man's characteristic gearshifts. Erinnerung is, for de Man, a repression of Gedächtnis, and the latter "is a truth of which the aesthetic is the defensive, ideological, and censored translation."[234] However, he also argues that, in the face of negative self-knowledge and the threat of self-erasure in thought, the defensive move contained in Erinnerung is necessary for avoiding total paralysis. In turn, memorization is given additional significance and de Man argues that it is Gedächtnis that plays a mediating role in the progression from perception through representation to thought. Memorization often makes use of material inscription; we write things down in order to forget them or we find ourselves copying a text mechanically. Thus, rote memory is a "machinelike exteriority" through which "the intellect, the mind, or the idea leaves a material trace upon the world."[235] Two things follow from this argument. Firstly, sign and symbol are not simply opposed categories, but, like the self-effacing I, exist in a relation of paralysis or "mutual obliteration."[236] Secondly, de Man argues that memorization can only be preserved in the form of its opposite:

> Memory effaces remembrance (or recollection) just as the I effaces itself. The faculty that enables thought to exist [Gedächtnis] also makes its preservation impossible. The art, the *techné,* of writing which cannot be separated from thought and from mem-

> orization can only be preserved in the figural mode of the symbol, the very model it has to do away with if it is to occur at all.[237]

Gedächtnis ensures the material inscription and sensory manifestation of the idea; we inscribe, so that forgetting is not total, but, he concludes, "the sign, random and singular at its first position, turns into symbol" or "the sign can only survive as a symbol."[238] De Man's closing discussion returns to the meditation on Giotto's allegories in *Swann's Way*—Proust's symbol that is not symbolic, the sign that can only survive as a symbol (and the "I" that must, but cannot, be enunciated), being allied with de Man's account of how Hegel's aesthetic "is no longer aesthetic." Hegel's "disarticulation of subject and predicate" in allegory is necessary for, yet simultaneously hinders, its meaning, as an allegory's specific predicates come into contradiction with their demand for abstract meaning.[239]

Benjamin's critique of the Neo-Kantians runs close to the one we have seen in de Man's late essays. We have already encountered the key passage in which he finds a type of synthesis in allegory that is invisible to the symbol's Neo-Kantian advocates, who perceive only allegory's disjunctions, and see them as only a drawback to the mode. It is a highly contingent "synthesis," to be sure, an open dialectic always liable to generate new fractures, a fragile suspension that lies, we should not forget, "not so much in the sense of a peace as a *treuga dei* between the conflicting opinions."[240] Interestingly, for Benjamin, the dialectical solution to allegory's antinomies is to be found "in the essence of writing itself."[241] In the more complex conceptions of allegory presented by Benjamin or de Man, the fact of antinomy and disjunction is displaced by a process internal to these contradictions where the act of inscription is significant. Commentators on Benjamin's *Trauerspiel* book sometimes distinguish the final section from the rest of the text, finding in it only an awkward break between the concluding redemptive argument and the main account of allegory. Yet, Benjamin's account of allegory's tentative synthesis occurs not, as might be expected, in the closing paragraphs on salvation, but in the opening sections of the chapter on "Allegory and Trauerspiel." Furthermore, he explicitly relates these two aspects of his work:

> For it is to misunderstand the allegorical entirely if we make a distinction between the store of images, in which this about-turn into salvation and redemption takes place, and that grim store

> which signifies death and damnation. For it is precisely visions of the frenzy of destruction, in which all earthly things collapse into a heap of ruins, which reveal the limit set upon allegorical contemplation, rather than its ideal quality.[242]

For Benjamin, allegory's destruction not only serves to challenge the ideology of symbolic consciousness, but also functions as the precondition for grasping allegory's own limits; this is fundamental if allegory itself is not to be reified. Acknowledging these "limits" is not to allegory's detriment; on the contrary, this perception, this "looking the negative in the face," which involves seeing a processual movement "beyond," is that which generates allegory's full effect.

Destruction, antinomy, disjunction, and rupture, then, are not the ideal qualities of allegory. Indeed, Benjamin argues that it is the symbol that idealizes destruction; allegory, in contrast, addresses "the *facies hippocratica* of history as a petrified, primordial landscape."[243] As de Man remarks in his essay on Hegel's *Aesthetics*, it is equally vain to deplore *or* praise the disjunction of subject and predicate.[244] Overdetermined by the critique of Greenberg's modernism, the antidialectical tenor of the cultural debates of the eighties too readily reduced these disputes to simplistic schemas that pitted allegory against symbol. Buchloh's survey of allegorical practices stands out from other accounts. Pursuing a more Brechtian perspective on Benjamin, he emphasizes the moment of allegorical reversal. Sherrie Levine's approach comes closest to allegory's melancholic and represents, Buchloh believes, "the strongest negation within the gallery framework."[245] Buchloh describes how the "allegorical mind sides with the object and protests against its devaluation to the status of a commodity by devaluating it a second time in allegorical practice."[246] Levine, he thinks, enacts this second devaluation, yet her "apparently radical denial of authorship" is in the end "complacent in defeat." Her strategy encourages a "fatalistic acceptance" and "a silent complacency in the face of the static conditions of reified existence."[247] The work does not open up, what he calls, "a dimension of critical negativity." Like Benjamin's melancholic, then, Levine's practice endures "the violence of the passive denial that the allegorical subject imposes upon itself as well as upon the objects of its choice."[248]

Although the reversal is already articulated in the *Trauerspiel*, Buchloh locates Benjamin's departure from this melancholic sense of allegory and into a politically inflected version with his work on Baudelaire and with the essay "The Author as Producer."[249] Producing art outside the accepted frames of reference and refusing to supply the apparatus, the work of Martha Rosler (see figure 3.4), for Buchloh, best approximates to this radical direction to allegorical praxis, although he expresses concern that her practice is subject to a certain level of impotence by way of this very externality. Here Buchloh keys into an argument that Rosler herself articulates in a trenchant appraisal of the methods of ironic quotation and appropriation then current. In this article, she deploys Lukács's critique of expressionism—where he addresses how fragmentation at the level of art reproduced fragmentation at the social level—to argue that contemporary appropriation techniques, despite the critical claims being advanced for them, end up replicating rather than undermining oppressiveness. Such strategies, she argues, remain too firmly tied and specifically referenced to their source material (the language of advertising or melodrama especially); simultaneously, while they adopt a totalizing critique that positions itself didactically above other art practices, their professed political target is too unspecific. As "there is no vantage point outside from which to make or understand the critique," appropriation, she argues, is impotent.[250] Buchloh's apprehension over Rosler's politicized approach not only seems prompted by the critical currency of appropriation practices (such as those of Levine and Prince, or by Sherman's "quotations" of female role types), but also foreshadows the air of social disillusion that he later develops (and which will be addressed in the next chapter).[251] As Rosler concedes, when pressed on the possibility of an external perspective on capitalism and its culture—of critical distance—Derrida admitted his support for the idea of social transformation; but in the climate of the early eighties, the overriding concern was shifting toward a critique not only of the old Left but of active politics and its supporting epistemologies.[252] (*October* also experienced internal division of a related nature when they declined to publish some presentations from an event that Crimp had organized on lesbian and gay media representation; ostensibly made on the grounds of quality, the disagreement over this editorial decision ended with Crimp's resignation.[253]) By now de rigueur for critical art, the "politics of representation" was often little more than a routine counterintuitive critique made against the "representation of politics." Certainly no less committed to the critical exploration of representational politics—indeed, more so—Rosler held firm to a political project both in and of representation, taking seriously the possibilities of its inner contestability. It is mark of how much has changed that the immediate ascendency of the model of appropriation adopted in the late seventies and early eighties should give way, already by the end of the century to a critical mood far more sympathetic to Rosler's position.

Among the authors associated with *October* at this time, probably only Buchloh—who began publishing in the journal from 1981—knew "Central Park." Still, there is a significant occlusion that speaks volumes of the intellectual mood: "Baudelaire's original interest in allegory," Benjamin notes, "is not linguistic but optical."[254] An emphasis such as this could not be addressed at the time because it compromised the challenge to the type of modernism advocated by Greenberg and Fried, which continued negatively to preoccupy *October*. (It is only relatively recently that Hal Foster has accepted that the Greenbergian position and its determinate rejection—which shaped the *October* project—no longer animates contemporary artistic debates.[255]) Strategically, the introduction of complexities, such as Benjamin's interest in the optical (and in the work of the nineteenth-century art historian who introduced the distinction between "haptic" and "optical," Aloïs Riegl), could only muddy the polemical waters, and it is intriguing to look back at this period to identify what is and what is not seen in Benjamin.[256] Other forces were at work in this transition between the Benjamin of the seventies (the Marxian theorist of antiauratic montage and reproductive technologies) and the poststructuralism of the eighties (especially in the particular form that Krauss and her students preferred).

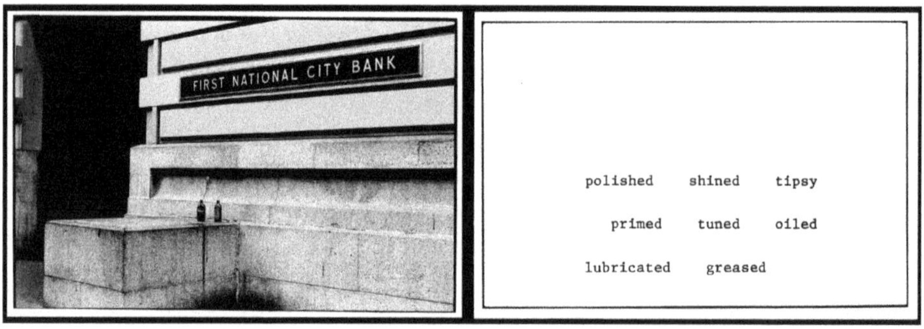

3.4 Martha Rosler, detail from *The Bowery in two inadequate descriptive systems*, 1974/1975. Copyright courtesy of the artist.

Putting *this* transition together with the project to counter the symbolic aspects associated with high modernism, it is easy to see how a certain intellectual narrative was written, one which—albeit now stripped of the explicit terminology of postmodernism—continues to be embedded in contemporary discourse.

Chapter 4

The Immobilizations of Social Abstraction

> . . . abstraction essentially is what rules Capitalism.
>
> —Robert Smithson[1]

> It is as ridiculous to yearn for a return to that original fullness as it is to believe that with this complete emptiness history has come to a standstill. The bourgeois viewpoint has never advanced beyond this antithesis between itself and this romantic viewpoint, and therefore the latter will accompany it as legitimate antithesis up to its blessed end.
>
> —Karl Marx[2]

Part I

In his defining essay "Postmodernism, or the Cultural Logic of Late Capitalism," written in the mid-nineteen-eighties, Fredric Jameson argued that critical distance "has very precisely been abolished in the new space of postmodernism."[3] Writing around the same time, Hal Foster noted how social difference becomes irrevocably problematized in the wake of "bourgeois divestiture." His conclusion is especially sweeping: if difference can be "fabricated in the interests of social control," he observed, "so too can resistance."[4] He reiterated the problem in another article:

> the real radicality is always capital's, for it not only effects the new symbolic forms by which we live but also destroys the old. More than any avant-garde, capital is the agent of transgression and shock—which is one reason why such strategies in art now seem as redundant as resistance seems futile.[5]

Symptoms of a growing consensus emerging in the eighties, these striking statements codified concerns that had been voiced for some time in art criticism and social theory. This discourse on the diminishing prospects for projects of social emancipation emerged prominently in discussions concerning the demise of the avant-garde.[6] Sometimes played out as a transformation taking place within the avant-gardes of the early twentieth century; at other times as the transition from the historical avant-gardes to their postwar neo-avant-garde successors; presented at other moments as characterizing the conversion from modernism to postmodernism: this account of a passage "from . . . to . . . " repeatedly charted the collapse of radical intentions and aspirations.[7] The debates around conceptual art provided another important strand to this storyline, and, within the space of just a few years in the early- to mid-seventies, condensed liberatory ambitions for an art resistant to commodification with the intense disappointment in the failure to realize these hopes, and with the idea—initially contested—of "dematerialisation."[8] Influenced by the sense of disillusionment following 1968, an analogous account of the loss of materiality took hold in the social and political theory with which many on the New Left engaged, a development closely associated with the theorization of de-industrialization and Post-Fordism, with the displacement of analyses of production for those of consumption, with the efforts to account for the impact of the mass media and advertising on postwar Western culture (and especially on ideology and consciousness), and with retreats from class analysis.[9]

Although its terms are increasingly questioned, the resulting synthesized social-aesthetic narrative of a "dematerialized" world is still with us. Drawing on Marx's account of the value form and the commodity's social and geopolitical universalization, this hegemonic narrative describes a loss of material conditions and coordinates; a process of social abstraction; an incorporation into the "culture industry," or into Enzenberger's "consciousness industry," or into the "spectacle," or into what is occasionally described—in an extension of Wolfgang Haug's discussion of postwar advertising techniques—as "commodity aesthetics."[10] Invoking the effects associated with modernity—increasing complexity of economic and social life; instrumentality, "disembodied" rationality, and runaway technology; the practical and institutional division of labor and the separation of "the value spheres" (science, ethics, and aesthetics)—"social abstraction" is widely understood as a process in which material characteristics evaporate into a flux of relations, where the commodity (as ware) gives way to money as the primary locus of worth; use value becomes subsumed by exchange value, and capital itself "dematerializes" into the fictions of financial capital. With these transitions, critical distance and practices of resistance, it is alleged, are terminally weakened, if not eradicated. Opposing chains of associations are established: on the one hand, the categories of production and labor are collapsed into "productivism" and aligned with "concrete positivity," physical materiality, and use value; on the other, exchange value is frequently hitched to consumption, abstraction, and, often by way of the linguistic analogy, with the Saussurean sign.

The argument just described has now become commonplace and is dominant among those accounts which aspire to articulate a social-aesthetic theory. The narrative of use and exchange value provides an important framework, for example, in Rosalind Krauss's book on Pablo Picasso (via a reading of Andre Gide's *The Counterfeiters*) and also for her discussion of Marcel Broodthaers's "eagle principle."[11] It is addressed directly in an essay dedicated to "value" in *Critical Terms for Art History*.[12] But these are late examples of an already widespread idée fixe. Looking back at the early nineteen-eighties, it is possible to detect a qualitative shift characterized by a strong fusion of the discourses on dematerialization and social abstraction into a synthesized social theory of art. Focusing on Benjamin Buchloh, Hal Foster, and Fredric Jameson, this chapter tracks how this meta-account was produced and, further, how it was developed and sustained by thinkers committed to a social and political analysis of culture. While it would be wrong to suppose that the work of such writers constitutes a homogenous thesis, it is significant that they develop their accounts in a sympathetic relation to the traditions of Critical Theory and Western Marxism. Typically, these theorists adopt a critical attitude to the cultural forms of late-twentieth-century capitalism while attempting to locate ways to outwit late-capitalist processes of social incorporation. These are positive and progressive aspirations. Nevertheless, the metaemplotment of social abstraction that shapes their projects needs to be questioned. Describing an epochal change in which social relations and experience are seen as increasingly abstract, this account implicitly proposes a socioaesthetic and political destiny—and one which has become part of the problem facing radical thought.

Buchloh's Shift

The presence of Buchloh and Foster in the orbit of *October* has been important for sustaining the more explicitly political elements in the journal's output. Initially contributors, their role became increasingly central, Buchloh taking on an advisory editorship late in 1986, Foster in mid-1990, with both becoming full editors in early 1991. Indeed, their influence here can be seen as contributing, without anyone intending, to the establishment of the synthesized socioaesthetic account under examination in this chapter.[13] Both critics came to prominence in a period when, in the midst of the "culture wars," the political stakes of art critique seemed especially high. It was necessary, as Foster articulated, to make a double challenge, not only confronting the paradigm of a high modernism that was still felt to be present, but also countering the "historicist" modes of the new postmodernism. The second challenge involved working against postmodernism's erasure of the conflictual dimensions of history, questioning its substitution of "false monuments" (to quote Buchloh) and opposing its naturalization of a neutered and closed historical narrative—the same antagonism to postmodern historicism that was shared, as we saw earlier, by the Revisions group. The project was to preserve critical and emancipatory space.

Formed in the context of the German New Left, and leaving for North America in response to Germany's political clampdown of the nineteen-seventies, Buchloh brought the emancipatory dimensions of Frankfurt School Critical Theory to the analysis of postwar artistic practice. His criticisms of Andy Warhol continued to show the traces of this intellectual formation, although he tried to distance himself from those tendencies that saw in more recent cultural production only a falling away from critical projects of the early twentieth-century avant-gardes. This refusal to dismiss the neo-avant-gardes out of hand did not result, of course, in a blanket embrace of all postwar art, and a key characteristic of Buchloh's approach has been his determination to distinguish between its progressive and regressive tendencies. Understanding these tendencies in both social and aesthetic terms, his critical process is not simply directed at separating out the most obvious targets (as, for example, in his critique of the conservative "returns" to painting in the eighties). He also discriminates within those more critical (or seemingly radical) art practices, including among those that challenged the Greenbergian paradigm (navigating, for example, the problem of the institutionalization and commodification of Minimalism and Conceptual Art).[14] Twentieth-century art is treated as a field where the specific social consequences of specific aesthetic strategies must be analyzed.

Buchloh is an essayist, and his essays are mostly monographic in type. Among the artists whose works draw his attention, there is a strong representation of artists with articulate political projects and an awareness of the historicity of aesthetic debates: Hans Haacke, Marcel Broodthaers, Martha Rosler, and more recently, Allan Sekula and Thomas Hirschhorn—practices often marginalized with the tag "political art." It is predictable enough that a figure like Buchloh should be attracted by the social commitment of these artists, but it is the conjunction of this political interest with a critical attention to form that seems especially important to him. Alongside these clear preferences are other artists whose critical credentials are less directly acknowledged and more contested, and whose case is made, or interrogated, by Buchloh; some choices seem more personal or contingent. Yet, whatever the orientation of his object of study, nearly every essay is an astute and probing exposition of a body of artistic work, in which Buchloh plunges into his topics as if seeking to seize hold of their political heart.

Neo-Avantgarde and Culture Industry brings together a number of his essays published between 1977 and 2000. In an effort to account for the development of his own project from the seventies to the nineties, Buchloh claims that, commencing with an essay on Hans Haacke from 1988, he moved away from an earlier, more explicitly New-Left perspective. In Buchloh's eyes, this change in his own work involved distancing himself from a paradigm that he had initially shared with writers such as Peter Bürger, namely, a conception of the relation between the avant-garde and neo-avant-garde in Oedipal terms and, secondly, an adherence to a mode of aesthetic judgment based on a "compulsory mode of critical negativity."[15] His dissatisfaction with Bürger's account of the neo-avant-garde typifies his revised relation toward the New Left. Buchloh sees his work as increasingly coming to focus on "the aesthetic capacity to construct the mnemonic experience as one of the few acts of resistance against the totality of spectacularization."[16] And, as shown by his essays on Haacke, James Coleman, David Lamelas, Yve Klein, and Arman, the framework that he develops describes art as caught in a tension between, on the one hand, reification and spectacle, and, on the other, efforts to preserve critical distance through the aesthetic enactment of historical memory. Indeed, Buchloh comes to locate his project methodologically as lying between *Dialectic of Enlightenment* and *The Society of the Spectacle.* [17] The change in outlook that he describes is not absolute; it is certainly no volte-face; nor does it represent some fundamental epistemological break. Nevertheless, his early writings do not always fits so neatly with the description he gives of typically Leftist approaches based on "mechanistic" ideology critique. These essays are not quite the embarrassment that Buchloh sometimes implies and his writing already displays some of those characteristic critical instincts that, retrospectively, he posits as the method that is yet to come into being, that is, one which understands "the multiple mediations taking place within each artistic proposition and its historical context."[18] Consistently and explicitly, Buchloh frames his project as a dialectical one. I will come back to the question of the differences between the "earlier" and "later" Buchloh, but for the moment we should note that the framework for many of his earlier essays is provided by the distinction between of use and exchange value.[19] The characteristics that Buchloh attributes to use value play a significant role in the formation of his intellectual project—and, by way of negation, in its subsequent transformation.

The relation of use value and exchange value, Buchloh argues in an article on Daniel Buren in 1981, lies at the heart of one of the "most crucial problems in twentieth-century art: the dialectic between aesthetic reification and the counter-concept of aesthetic use value."[20] Through his essays, this "dialectic" is variously described as dilemma, schism, double bind, tension, and ambiguity. Buchloh describes Michael Asher as working on the threshold of "symbolic" and "actual" space, "continuously increasing the ambiguity between functional object and aesthetic object, as though to prove from within the analysis of sculpture itself that it has lost its material and historical legitimacy."[21] If not at an absolute standstill, this is a dialectic that is rapidly atrophying, expending its energies circling within, or ricocheting back-and-forth between, its defining terms, its movements being those of exacerbation and hypostatization. Fragmentation takes the form of a fundamental contradiction, and Buchloh often uses allegorical structures to describe its paralyzing fractures. Buren's 1977 project, *Les couleurs: sculptures* (see figure 4.1), is interpreted by Buchloh not just as dialectical form, but as "a *grotesque* in the true sense of the word."[22]

Buchloh's account of the dialectic assumes slightly different forms. The schism of "aesthetic reification" and "aesthetic use value" is sometimes rendered more simplistically as one between "aesthetic object" and "functional object"; or as the aesthetic sign versus real objects (or processes); or, more simply still, as a contrast between sign and function.

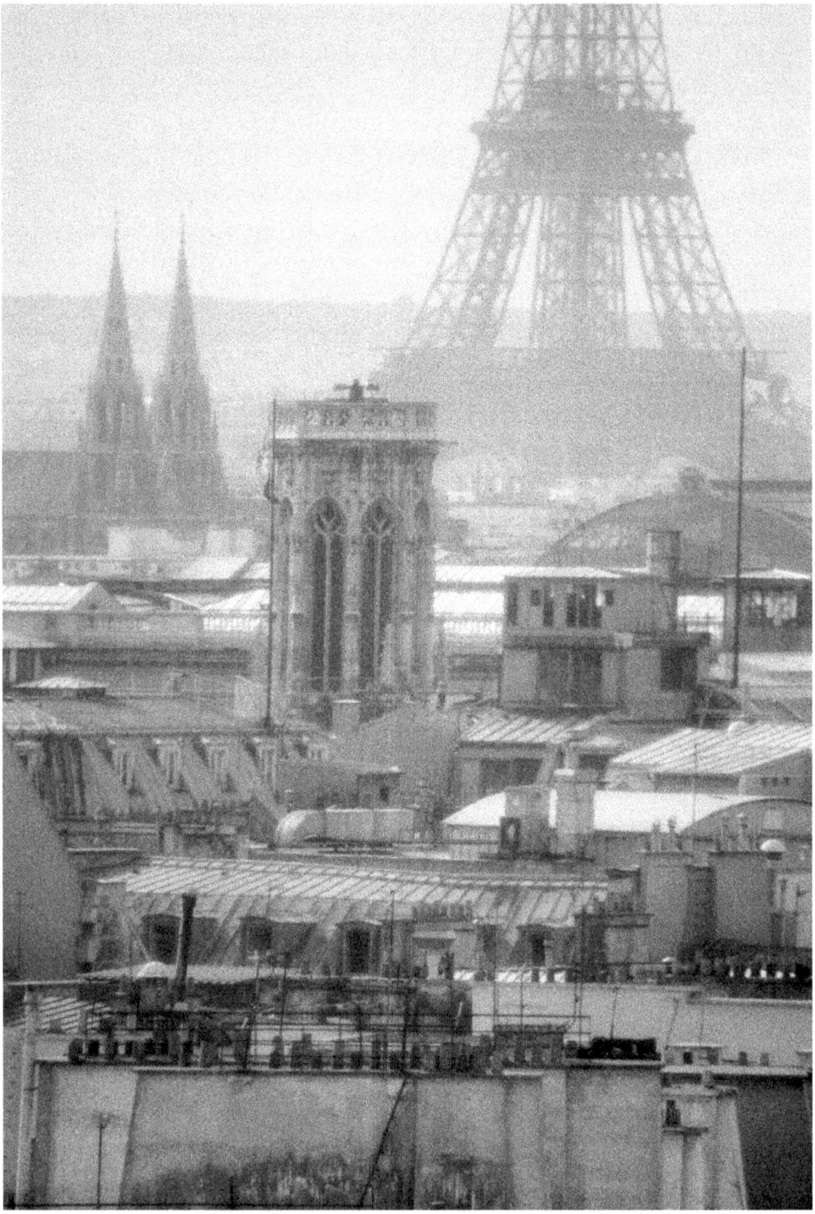

4.1 Daniel Buren, *Les couleurs: sculptures*, 1977, photographic record of detail of work *in situ*, MNAM, Centre Georges Pompidou, Paris. © ADAGP, Paris and DACS, London 2009.

Clearly, the implications of these oppositions are not necessarily synonymous, although it would seem that Buchloh is attempting to find different ways of saying much the same thing. But we should note that while the first of the tensions just cited makes an opposition *within* the aesthetic, the second externalizes aesthetic qualities to one side of the dialectic's contradiction. In the wider context of Buchloh's essays from this earlier period, it is the latter version that dominates, and the aesthetic is identified with reification. In his later writing, it is the aesthetic itself that is treated as riven, as a process of reification fully embedded within the commodity and yet also endowed with mnemonic critical potential. Yet, it is important not to forget that the outlines of the later work are already present in earlier essays. Even in the pre-1988 essays, where the contrast of aesthetic and function is still the language of choice, the substance of Buchloh's discussion of specific practices often subjects his material to a further reflection. Thus, Buchloh describes how while Buren's *Les couleurs: sculptures* comprises functional and aesthetic elements, it is the signs (flags) that are "functionalized" and the functional objects (telescopes) that become "aestheticized."

With these terms, Buchloh develops a more general account of contemporary art. Art's increasing disjuncture from reality is determined by the changing political situation, and this unshackling of art's role hypostatizes the opposition of sign and function into the "arbitrary sign" and "obedient object":

> Inasmuch as any work of art becomes increasingly superfluous under the conditions of total reification because it has lost its function as a model of critical reflection of social reality, it approaches a state of either mere objecthood or of mere aesthetic voluntarism, i.e., decoration.[23]

Here the stakes that Buchloh invests in the category "use value" become clearer. The functionality to which he repeatedly returns does not simply refer to art's usability as an object in the narrowly literal sense; materiality, as Buchloh makes clear on a number of occasions, has to be understood as a social category (none of which, per se, prevents his awareness of the play of literality and facticity in postwar art).[24] Its use as an object for aesthetic engagement is understood not as aesthetic enjoyment but rather as a model of social critique, or, as he also puts it, "its essential capacity for negation."[25] Yet, as Buchloh notes in an article from 1982, in the postwar period especially, this critical capacity increasingly comes under threat, to such an extent that the artist is "condemned to produce pure exchange value."[26] Aesthetic validity in contemporary art is determined not by the capacity for negation but, as he puts it, by the "capacity to generate exchange value."[27] Buchloh alludes to Minimalism's appropriation into the nexus of the commodity and the collapse of its critical dimensions, its objects "acquiring exchange value inasmuch as . . . their context-bound idea of use value is abandoned."[28] In his discussion of montage techniques, this problem is applied retroactively to the historical avant-garde and he notes how their objects "ultimately function exclusively as producers of exchange value."[29] As he puts it in 1978, "if artistic production gives up altogether the idea of use value, it abolishes its own inherent potential to induce dialectics within the reality of cultural history, thus producing mere artistic facticity incapable of initiating further processes of development."[30] It is the artists whose projects Buchloh considers to provide resistances to this situation, and who refuse to give in to "the general conditions of production"—artists such as Haacke, Asher, Buren, Dan Graham, Lawrence Wiener—who draw his critical support: Graham attempts "to reinvest the artwork with a potential use value"; Haacke believes that "art should have a use value" and provides a critique of "art as exchange value."[31]

In his earlier essays, Buchloh also charts two main avant-garde legacies, that of Marcel Duchamp and that of Soviet productivism. Although he actually detects an original productivist dimension to Duchamp's intervention (the readymade, he argues, was inserted into the aesthetic field as an object of use), he believes this aspect gave way to the idea of the readymade as the locus of philosophical speculation on art; thus, Duchamp comes to represent "the symbolic substitution of use value objects for exhibition/exchange value."[32] In 1978, and in a more implicit comment, Buchloh notes how Dan Graham's materials comprise not found, readymade objects, but instead "found structures beyond visible reality and its seeming concreteness."[33] This statement echoes that of Walter Benjamin, Bertolt Brecht, and Fritz Sternberg on the inability of photographs of factories to tell the viewer anything about the social relations of capitalist industry. The strategy of *Verfremdungseffekt* or *ostranenie*, of Brechtian "artificial construction" or Shklovskian "knights' move," then, is counterposed to Duchamp's insertion of found objects of mass production into the aesthetic realm (rather than, as in the example of the photograph of the factory, being contrasted to the illusory immediacy of appearances assumed by uncritical reportage). In a 1988 essay, Buchloh argues that it was the Duchampian model—with its revivals in the fifties and sixties (by John Cage, or through the art of Pop and Fluxus)—that artists such as Broodthaers, Buren, Asher, Smithson, and Haacke challenge. Here, Buchloh's argument is clearly with a further reenergizing of neo-Duchampian strains: Duchamp's mediation in the eighties by a popular appropriation of Baudrillard's theory of the simulacrum. In this context, Haacke's *Manet-Projekt* of 1974 is deployed by Buchloh as an alternative model. However, the resistive quality of Haacke's work is seen not as lying in a simple counterassertion that elevates the qualities associated with use value over those of exchange value (the latter, for Buchloh, are characterized by the abolition of referentiality, context, and specificity, and by the establishment of the rule of equivalence). What gives *Manet-Projekt* its critical charge, Buchloh argues, is that Haacke takes the dominant narrative of exchange value as his very material, into which he enfolds a strategy of "critical negation," emphasizing functionality as well as the specificity of site, time, and object, and "by linking allegorical strategies to instrumentalizing acts of information and communication."[34] In Buchloh's view, Haacke does not seek to evade the rule of exchange, nor does he attempt to reclaim some version of function and material specificity untrammeled by the commodity; rather, he addresses contemporaneity head-on, by taking its givens as a historical problem to be critically interrogated. Buchloh notes how the artist provides an allegorical practice in which "devaluations of the aesthetic object" are conducted "within the object itself."[35]

Thus, the project of reinstating use value is contrasted with other responses to the historical dilemma facing modern art. As Buchloh points out, "Use value is art's most heteronomous counterpart."[36] One of these reactive responses tries to recover the stability and materiality of an older aesthetic—in Buchloh's terms, to recover the qualities of art before exchange value. Such artists might reach back for materials and procedures grounded in tradition, but far from recovering historical value, their work merely erects false "monuments" which reinforce social amnesia.[37] Other approaches that Buchloh has in mind take a more critical stance by attempting to resuscitate the radical strategies of the avant-gardes of the twenties. However, he argues, productivism's artistic strategies cannot be repeated, at least, not without transforming their content into "the opposite of their original intentions" and reducing them to "formal surfaces." Therefore, they can only achieve "the opposite of their original intentions." As an "instrument of materialist recognition and transformation," use value in Buchloh's writing must be understood as a category that is bound by its social context.[38] Consequently, belated efforts to return to the modes of the Soviet Constructivists end up reemphasizing the wider evacuation of social and critical content. The specificity that Buchloh takes to be so crucial to use value and materiality concerns the social particularity of a historical moment; this important desideratum, therefore, now means that we come to terms with the conditions of the postwar culture industry and, from the seventies, also necessitates recognizing the terms of art's postconceptual problematic. However, if many of the efforts to recover use value—or, at the very least, its potential—present artists with historical difficulties, it is, as Buchloh indicates, scarcely any easier for neo-Duchampian strategies. Each act of appropriation, far from achieving the intended "interference within the signifying practice," succeeds only in widening the very gap it seeks to cross, turning "into a renewed legitimation of existing power structures."[39] This gambit (one especially associated with the neo-avant-garde but also pertinent to the art of the eighties) also reaches an historical hiatus. As Buchloh later clarifies, he comes to identify a sea change around 1975, after which, in the postconceptual climate, the legacies of the historical avant-garde cease to animate contemporary practice in the ways still possible between 1955 and 1975.

> I would suggest that only at this time [the midseventies] did a radically different basis for critical intervention in the discursive and institutional frameworks determining the production and the reception of contemporary art become established, generating propositions of audience reception, distribution form, and institutional critique that were distinctly different from the critical models invoked by Bürger.[40]

Buchloh, therefore, closes this phase of the neo-avant-garde; Haacke, Buren, Broodthaers, Asher, and Graham point the way to its successor.

At first sight, and in the context of Buchloh's comments, the project to reassert the potentials of use value seems to be part of the New Left discourse from which Buchloh claims to depart in the late eighties. Just as positive allusions to productivism cease to provide a point of reference, so too does "use value" disappear from Buchloh's terminology. However, on closer inspection, things are not so straightforward. The category of "reification" has staying power, as does "exchange value"—or, as Buchloh sometimes puts it, "sign exchange value"—and he draws frequently on "spectacle." These terms provide convenient conceptual shelters under which can be gathered all those qualities that insinuate the (dire) state of the contemporary world. In other words, the displaced term—"use-value"—continues to provide the determining structure for, and social theory delineating, his thought; it is just that it now hovers off-screen as the implied conceptual other, figuring the past . Interestingly, just as the explicit discussion of exchange value continues in his writing, so his critique of the Duchampian legacy persists (this can probably be put down to its continued currency within the discourses of contemporary art); if anything, Buchloh's attitude toward this legacy becomes more trenchant. If the quality of exchange characterizes the readymade, Buchloh further argues that, in extending the Duchampian project, John Cage stretches the domain of equivalence to include the exchangeability of creative modes (music, text, gesture, as well as object). Both Duchamp and Cage, Buchloh continues, "embrace reification outright," a tendency that culminated in Pop's "aesthetics of indifference and of pure affirmation."[41] Buchloh, then, presents Pop and tendencies of Fluxus as engaging in a reifying aesthetics, contributing to a culture in which art becomes fully identical with "the economic order of exchange value produced by commodities" and which "naturalizes the apparent inescapability of an ever-increasing totality of reification and of fetishization in a strategy of cool and affirmative dandyism."[42]

Writing in 1997, Buchloh summarizes a further characteristic of the post-neo-avant-garde period:

> the avant-garde's former claim to disregard the persistence of social memory embedded in institutional practices and discursive conventions has been exchanged for the recognition that it is precisely from the reflection of these conventions that the only

> remaining potentials for resistance against spectacle culture and the totality of administrative control can be mobilized.[43]

Memory increasingly comes to the fore in his essays in the nineties. Indeed, it takes on the role once performed in his account by use value: the resistance to commodity culture. It is worth noting how Buchloh uses the category of memory. He distinguishes "historical memory" from "discursive memory," the latter being a narrowly formal or procedural version of recollection that recalls past genres or methods.[44] Thus, even memory itself is riven in Buchloh's schema by the same fundamental schism. So while the avant-garde (and neo-avant-garde) prohibition on the past disappears from advanced artistic practices, art after the midseventies is still subject to the same problems of historicity. But the situation he describes is even more desperate than it first appears: memory provides resistance's last flickering of potential. Although Buchloh ceases in his later writing to present a stand off between "aesthetic" and "functional" practices, and in this sense develop a more complex resource for thinking about art as a critical practice, he simultaneously loses the motor for critical release that was involved in his earlier discussions. This release did not come from his conceptualization of functional practices *per se*, nor did it derive from his hostility to "aesthetics." Buchloh's changing attitude certainly mirrors the political disappointments of a generation, but it also has a conceptual basis: the presence of both use and exchange value in his earlier writing sparks the crucial discharge. However, his conceptualization of these categories is problematic; much in the manner of the Frankfurt School, he generalizes Marx's distinctions into a totalizing account of reality, and his conceptualization of the stand off between materiality and sign, as well as past and present, is just too undialectical. Nevertheless, despite its limited conceptualization, his earlier essays manage to sustain some dialectical momentum simply by having use value in play. Buchloh's may not be an intellectual project that naturalizes the inescapability of deepening reification—he may even point to the very apparentness of *this* trap—but increasingly his essays posit all but absolute reified relations in the realm of the culture-industry-turned-spectacle, and he struggles to see much beyond the straightjacket of a world shaped by the logics of "sign exchange value." The culmination of this trajectory is evident in his contribution to a roundtable discussion on "the predicament of contemporary art," where, with Hal Foster repeatedly trying to pull him back from the brink of the defeatist abyss, Buchloh's assessment assumes critical closure.[45]

Foster's Holdout

As an ambitious young writer on art, Hal Foster shaped a project of critical intervention in the context of the culture wars in the United States; the still keenly felt legacy of Greenberg's modernism; reactions to the conceptual and politicized work of the seventies (the return to painting, the "new spirit," etc.); and the emergence of a generation of young "postmodern" artists on the New York scene (especially those associated with Crimp's exhibition and article "Pictures"). In a series of articles written in the nineteen-eighties for art journals, especially *Art in America*, in his collection *Recodings* (1985) and in his anthology on postmodern culture (1983), Foster articulates a social analysis of contemporary culture conveyed with a tone of political urgency.[46] Informed by various political and critical theories, Foster demonstrates a familiarity with the writings of Adorno and Gramsci, engaging with recently translated material such as Debord's *Society of the Spectacle*. It is in his essays from this period that Foster distinguishes two types of postmodern discourse. On the one hand, he identifies a neoconservative formation deploying a populist rhetoric in its opposition to modernism. This "postmodernism of reaction," as he calls it, aimed to recover "tradition" and "history" through the exploitation of codes, styles, and symbols, but at the cost of erasing or concealing social and political factors.[47] Foster describes the resulting pastiche as a dialectical reversal, where the very effort to retrieve style and history in the face of contemporary social fragmentation ended in the reinforcement of this disintegration, neoconservative postmodernism being just another symptom of the condition which it purported to flee. In contrast, he characterizes the second tendency of postmodernism as oppositional and as an explicit challenge to both the official modernism of the late twentieth century and the neoconservative postmodern reaction. Sometimes called "a postmodernism of resistance," Foster also dubs this tendency as "poststructuralist" or "critical postmodernism."[48] Crucially, rather than attempting to reinstate historical meaning, this tendency is seen as exploring the political values of textuality, questioning, and deconstructing cultural codes. Supporting the work of emerging artists such as Louise Lawler, Sylvia Kolbowski, and Sherrie Levine—contrasting unfavorably that of Julian Schnabel and Sandro Chia—Foster's work became closely identified with this theoretically informed opposition to neoconservative modes.

A formative voice in debates on contemporary art as a radical practice, Foster seeks out new strategies of cultural contestation. The influential collection *The Anti-Aesthetic*, from 1983, explores "practices of resistance." Elsewhere he considers how to devise an art that resists the cultural incursions of capital.[49] The problem of critical distance remains at the heart of his project; Foster addresses the debate over critique's permanently threatened state under conditions of increased social reification and the very question of what is meant by critical practice. Describing the compromised conditions of radical art practice in the early and middle nineteen-eighties, Foster expresses his fear that appropriation by capital "is so efficacious because it proceeds by abstraction."[50] For Foster, this process of abstraction and equivalence operates by acknowledging otherness while simultaneously reducing that otherness to sameness, and he cites Adorno and Horkheimer's comment that: "Bourgeois society . . . makes the dissimilar comparable by reducing it to abstract quantities."[51] From here, analogies are rapidly made between the declining potential of radical art, on the one hand, and, on the other, the reign of exchange value.

As perhaps indicated by the retitling of *The Anti-Aesthetic* as *Postmodern Culture* for the British edition, Foster became the voice of a radicalized brand of postmodern discourse that tried to sustain a spirit of resistance.[52] Nevertheless, in the climate of political defeat in the nineteen-eighties—the rise of neoconservative agendas (such as those initiated by Reagan and Thatcher) and the introduction of neoliberal economic policies—and at a point where poststructuralist discourse was in ascendancy in the Anglophone academy, Foster's essays also delimit the ambitions of critical practice. Advocating "practices of resistance" over "transgressive acts," Foster takes his distance on older models of avant-gardism (and, by implication, the politics of a revolutionary Left).

Foster's relation to critical postmodernism and poststructuralism, however, is less straightforward than it might initially appear. In "(Post) Modern Polemics a rather different view starts to emerge, and we get a sense of how Foster is torn between competing models of criticism. Having set up the contrast between neoconservative and critical postmodernism, Foster proceeds to articulate their similarity. Both tendencies, he argues, are "symptoms of the same 'schizophrenic' collapse of the subject and of historical narrativity—as signs of the same process of reification and fragmentation under late capitalism."[53] In both cases, the subject becomes decentered, history is eroded, and representation is shorn from any referent. The two tendencies may be opposed by politics and style, but, according to Foster, they are historically congruent. He notes how, for instance, poststructuralism loses critical power when it becomes a "rhetorical exercise without repercussive effect." It is found equally wanting for its treatment of modernism as a monolithic entity, and for reducing its characteristics to autonomy and self-referentiality.[54] Interestingly, however, Foster does not follow this argument through, and in the conclusion to the essay he takes back his critical remarks on poststructuralist postmodernism. This time around he reduces the symptoms of late-capitalist dispersal to Schnabel and Salle, and declares the demise of the bourgeois patriarchal subject to be "no loss." While one might be prepared to agree on this last point, this is a conclusion that performs a complete about-turn in his argument. Failing to address the central critical direction taken by his essay—by way of evasion and nonsequitur—this final move can only be read either as an overly hurried wrap up or as a failure of nerve. It appears that, in the climate of radical cultural debate in the mideighties, a sustained critique of poststructuralism risked compromising one's critical credentials. It is also noticeable that as "poststructuralist postmodernism" itself waned, Foster has allowed some of his unease to resurface. He sees, for example, both modernist and postmodernist discourses as threatening critical distance, and he also draws his readers' attention to the problem of poststructuralism's "epistemological exoticisms."[55] . In order to avoid the rhetorics of social distinction and colonialist conceptions of othering, and following the writing of Franz Fanon and Benjamin, he argues that critical distance must be reproposed as active and useful.

Despite some shifts in emphasis occasioned by developments in the nineties—the fall from prominence of the discourse of a critical postmodernism, and the banalizing of the term "postmodernism" itself—Foster's approach is remarkably consistent. There are repeated patterns to his thinking. Modernism did not fail but was absorbed into official culture (a culture that was dominant but dead), Foster insisted in 1983; and in more recent assessment, he observes that critical postmodernism did not fail, it just became unfashionable. While he touches on an important social phenomenon, Foster here employs an account that comes very close to the most traditional explicatory devices of art history: that of stylistic exhaustion and Oedipal reaction. As a force to assimilate critical cultural formations, the commodity logics of fashion are seen as displacing hegemonic culture. Another way of seeing this would be to say that official culture now finds its very constitution transformed; its power now residing not so much in a residual public sphere, framed by a high-cultural sector of the leisure industry, than in the ability of a commodity to sustain market success. This change of emphasis is accompanied by a loss of cognitive force; passivity and inescapability now seem bestowed upon the process; the assimilating power is no longer socially definable—a person, a group, a class faction, an institution—but is posed instead as an impersonal abstract. Commodities are social, of course, as are the processes of abstraction their production initiates, but, in this elevation of the commodity's power, the very essence of the category "social" becomes so etiolated as to make commodification a near-metaphysical state.

More recently, Foster's concern with supporting practices of resistance has translated into an account of how we might resist, what he calls, the "presentist totality of design in culture today."[56] The work of architect Rem Koolhaas takes on a central role in this recent discussion. Unlike Leon Krier's reactionary historicism, the commercial populism promoted by Robert Venturi, or postmodernism's pop-historicist compromise, Koolhaas does not abandon the modernist project. Instead, Foster argues, Koolhaas seeks to "recharge" modernism, "reprogramming" it along the lines of the "paranoid-critical method" once advocated by Salvador Dali.[57] Koolhaas, in Foster's view, also manages to reconcile the Bretonian and Corbusian strains of modernism, fusing irrationalist and rationalist tendencies (in this regards, Koolhaas is compared to the postwar avant-gardes and the Situationist International), and tries "to ride the dialectic of modernization in a way that might keep these projects alive for the future."[58] Foster also recognizes the "ambiguity" or "ambivalence" in Koolhaas's project. As he puts it, Koolhaas's efforts to "surf the dialectic" of modernization or the market are caught between critique and complicity, or between bravado and desperation. While not akin to co-option, Koolhaas' alleged "cynicism" and "surrender"—charges brought against him by his critics—is formed by the fact that (in an echo of Manfredo Tafuri and Massimo Cacciari) "architecture must attend to the *Groszstadt*, if not surf it." (Tafuri is one of those critical of Koolhaas's cynicism.) Foster immediately draws out the consequences: "it is difficult," he observes, "to imagine a politics today that does not negotiate the market somehow."[59] Complicity is somehow *necessary*. Foster sees this strategy as analogous to the "left-Fordism" of Benjamin and Gramsci, but where theirs was "a go-for-broke strategy that looks through and beyond this mode of production," the contemporary ambivalence of Koolhaas and others may make the strategy more myopic: it can produce critique and great poetry, Foster notes, "but that may be all."[60]

How, then, can we live on through the contemporary world's "ambiguous stay of sentence," the "lack of certainty" produced by the "weakening" of paradigms, he asks.[61] In this "paradigm of no paradigm," determinate responses—such as Judd's or Morris's reactions in the midsixties to the high modernism of Greenberg and Fried—are unable to thrive. In a nod toward Greenberg's early analysis of cultural crisis, Foster suggests that this situation might represent a "new Alexandrianism" (but nevertheless might be a "good" valorization and diversification of a "weak theory"). What is it, Foster wonders, to make art when even the postsixties "expanded field" has imploded? How do we negotiate a situation where the compounds once outlined in Rosalind Krauss's semiotic square have lost the capacity to provide productive tensions and, thus, a field's critical animation?[62] We find ourselves beyond the "post" and beyond the "neo," beyond the very distinctions between avant-garde and neo-avant-garde, as well as those between modern and postmodern: "the model of a formalist modernism challenged by an expansive postmodernism no longer drives or describes significant developments in art or criticism."[63] Foster also notes that contemporary artists are no longer concerned with the dichotomy of formalism and social approaches, instead making work that is both formally and socially astute. The same point is cited in his discussion of Koolhaas: it is important, he writes, "to reinvent a plausible relationship between the formal and the social."[64] It is not a return to grand narratives that is called for, but rather a need for "situated stories" which might "focus" practice. Foster sees himself as relatively optimistic (Buchloh being positioned as pessimism's representative), but his own prognosis is profoundly melancholic. We need to find a way of "living-on," he writes, drawing on Adorno and Beckett; this is about "making-do" with the situation, beginning again, and "making-new." Back in the eighties, Foster took his distance from Adorno's "negative commitment" and, instead, advocated a Gramscian "strategy of interference." Two decades later, Gramsci is figured as a Leftist version of completed nihilism, and Adorno represents a measured operation of holding out in our "ambivalent stay of sentence." Never mind that the names have switched places or that the values attached to Gramsci have been revised—a voluntaristic summons to the barricades may be inappropriate, but on both occasions, Foster's account performs the same function: despite his intention, and despite his playing the optimist to Buchloh's pessimist, this argument again underscores the impossibility of ambitions for radical social transformation.

Jameson's Loss

Like Foster and Buchloh, Fredric Jameson reflects on the fate of emancipatory thought under conditions of its erosion. A prolific writer—whose writings have expanded from their literary base to address a range of cultural phenomenon, from architecture to film and mass culture—Jameson provides the most influential of these synthesizing accounts. At first addressing the tradition of Western Marxism in *Marxism and Form*, in the seventies he went on to consider the relation between postwar structuralism and linguistic formalism as developed in Eastern Europe in the early part of the twentieth century. His ability to move between—and, sometimes, to meld together—the Hegelian-Marxist critical tradition, psychoanalysis, semiotics, and narratology distinguishes his work from those more inclined to proselytize on behalf of one or the other. Although his preferences are explicit—and this includes, against the grain, a continued positive assessment of the work of Lukács and Sartre—Jameson approaches these different theoretical trajectories in ways that are both critical of, and sympathetic to, his subjects. This suppleness, which has allowed him to speak simultaneously to distinct academic constituencies, is a defining feature of Jameson's intellect and a crucial source of his impact; and, as we will see, it produces tensions.

Above all, however, it has been Jameson's considerations of postmodernism from the early nineteen-eighties—vehicles through which he has made larger claims on the analysis of the culture of contemporary capitalism—that did most to consolidate his influential role in the humanities. In 1983, the year before *New Left Review* published "Postmodernism, or the cultural logic of late capitalism," Foster included Jameson's essay "Postmodernism and Consumer Society" in *The Anti-*

Aesthetic. Foster has maintained a close, albeit critical, relationship to the writings of Jameson, and among the theoretical approaches that Foster assembles for his arguments, it is noticeable how Jameson is often deployed as a corrective to the poststructuralist accounts. Jameson is also used to articulate relations between cultural analysis and historical developments within capitalism. (And it is interesting to observe that, in a 1996 essay, it is in a critical engagement with Jameson's account of postmodernism that Foster pursues the question of whether the very term "postmodernism" retains any critical power.[65])

Jameson's intellectual project takes the form of an explicit defense, and renewal, of Marxism, in a context of political defeat. In addition to *Marxism and Form*—a study of dialectical aesthetics in the work of Benjamin, Adorno, Sartre, and others—Jameson provides the extended afterword to the collection *Aesthetics and Politics.* He continues to rework the core issues from these debates: how the aesthetic and cultural forms of modernism relate to the commodity and its associated quantification and equivalence. In his writings on postmodernism, Jameson pursues this project with regards to the aesthetic and cultural forms of late capitalism. As the title to his famous essay from 1984 makes explicit, he had understood postmodernism as the cultural manifestations of a new form of capitalism, indeed, as its "cultural logic." Basing his analysis on Ernest Mandel's eponymous book of 1972, Jameson understands "late capitalism" as the third, and most recent, stage in capitalism's development. Late capitalism, then, is the successor to two earlier phases: mid-nineteenth-century market capitalism (associated in particular with the surge in development of productive forces initiated by steam technology) and the phase of monopoly capitalism set in train in the eighteen-nineties (characterized by electrical power and the combustion engine). Jameson identifies distinctive cultural superstructures for these two earlier periods, with nineteenth-century realism being supplanted by modernism. In his account, the emergence of multinational capitalism in the nineteen-forties—with the transition to nuclear and electronic technologies, and the extension of the media and advertising industries—inaugurates the postmodern cultural forms of the late twentieth century. Central to this argument are the key themes under discussion: the incorporation of the avant-garde and the end of negation; the subsumption of use value; and the rise of the processes of dematerialization and abstraction. Jameson treats late capitalism as an intensified or "purer" stage of capitalism, in which the forces of commodification have widened their embrace of, and deepened their penetration into, social and psychic life through the colonization of Nature and the Unconscious.[66]

It is around these questions that Foster's differences with Jameson come to the fore. Both Jameson and Foster regard cultural phenomena as symptomatic of, and responses to, changes in socioeconomic conditions, while being simultaneously equivocal about "reductive" interpretations of the "base-superstructure" model.[67] Nevertheless, Foster is critical of Jameson's—and Mandel's—"depth model" and he distances his own approach from the seemingly directive and historicist implications of Jameson's interest in cultural logics. For Foster, Jameson's account is too totalizing, "too spatial," and insensitive to history's heterogeneity.[68] Foster, on the other hand, emphasizes the temporal unevenness of capitalism's expansion, which, notably, he chooses to describe not through Leon Trotsky's important work on combined and uneven development but through the psychoanalytic concept of deferral, and specifically through Sigmund Freud's concept of *Nachträglichkeit*. Consequently, Foster—who consistently challenges those arguments that present postmodernism as an absolute break with modernity—takes postmodernism not to be a transition within capitalism, but rather as a phenomenon existing in "parallax" with modernism.[69] (In contrast to Foster and Jameson, the very term "postmodern" is conspicuous by its absence from Buchloh's writing, and on those rare occasions when he does deploy it, he clearly does so under duress.) Jameson himself acknowledges that he has an ambivalent relation toward the term "postmodern." As he puts it in his introduction to his 1991 book *Postmodernism or, the Cultural Logic of Late Capitalism*:

> The concept [postmodernism] is not merely contested, it is also internally conflicted and contradictory. I will argue that, for good or ill, we cannot *not* use it. But my argument should also be taken to imply that every time it is used, we are under the obligation to rehearse those inner contradictions and to stage those representational inconsistencies and dilemmas . . . . Postmodernism is not something we can settle once and for all and then use with a clear conscience.[70]

Needless to say, few heed these protocols. In *A Singular Modernity*, a similar argument is applied to "modern": "you must simultaneously affirm the existence of the object while denying the relevance of the term that designates that existence."[71] Nevertheless, in practice Jameson has continued to affirm and to provide detailed topographies of these problematic categories for designating their cultural logics.

Jameson proffers a complex range of examples as evidence for this epochal change from modernity to postmodernity ranging from the integration of art into commodity production through to the supposed demise of critical distance. In this new social formation, he argues, culture is dominated by the production and reproduction of simulacra. The signifier itself takes on heightened materiality at the expense of the signified; yet the signifier's materiality is unshackled and only exists among the "rubble" of other signifiers, each unrelated to any other. (This, we will recognize, is the "breakdown of the signifying chain" that Jameson reworks from Jacques Lacan's analysis of schizophrenia.) Under late capitalism, the subject is also transformed, and Jameson discusses the demise of the psychopathologies of the ego, such as notions of individuality and expression, as well as anxiety and the sense of alienation. The older modern ego is replaced by free-floating, decentered, fragmentary "nonselves" who are shorn of the capacity for self-reference, a process which Jameson presents as one of "self-annihilation." This new postmodern subject is a "spectral" character, whose scarcely perceptible, ethereal materiality or strangely disembodied embodiment remains a key figure in Jameson's later writings. It is important to note the contradictory doubling Jameson gives to this disposition where the immaterial is presented not merely as opposing materiality but primarily *as* a new material condition.

Jameson explicitly discusses the "waning of affect" and the "waning of historicity," but also presents us with the fading of a number of other features, each of which depends on, or has reference to, a "center" or "ground" including those around which the bourgeois subject orientated, such as selfhood, alienation, and those, such as the referent or the signified, which might seem to be guarantors of meaning.[72] Jameson's use of a language of "loss" should also be noted. As Jameson knows only too well, and as the older discussions of alienation show—discussions that range across theological and psychological, social and political, romantic, reactionary, liberal, or Marxist approaches—the discourses of modernity are riddled with such narratives: the loss of God, of unity, certainty, value, and meaning. (We might think here of the debates associated with Sartre and Simone de Beauvoir, Jürgen Habermas, and, more recently, the work of Agnes Heller and Simon Critchley, and the whole tradition associated with *Kulturkritik*, including, in its more conservative modes, Oswald Spengler's fears over the end of civilization and the decline of the West). So, for Jameson, it is important to mark a distance from these established accounts of modernity. Although he narrates loss at a metahistorical level, when he comes to outline the symptoms of late-capitalist culture, he describes a situation in which we have "lost the conditions of loss." Since "loss" implies the loss of something, a consciousness of what it is that is missing, in postmodernity it is the very idea of "loss" that has disappeared; we no longer have the ability to locate that loss, temporally or spatially; negation itself withers, lacking determinates to resist. In grasping Jameson's point, it also seems important to recognize the shift from singularity to plurality, or better, to think this transformation as a move from *the* singular to *a* plurality. Somewhere in all this dispersal into an aleatory and intoxicated heterogeneity, the pluralizing has to involve not only a quantitative multiplication but also a qualitative process of becoming indefinite. These arguments provide the (non)coordinates of Jameson's presentation of abstraction.

Contemporary capitalism is faced with the problem of historical thought. "It is safest," Jameson argues, "to grasp the concept of the postmodern as an attempt to think the present historically in an age that has forgotten to think historically in the first place."[73] Foster's critique of neoconservative postmodernism in the eighties—and, later, his conviction in the importance of restoring the "mnemonic dimension" to contemporary art—is concerned with the same issues of historical memory and narrative, the same problems that come to preoccupy Buchloh.[74] Jameson's narrative of the "loss of loss" is particularly important as we move on to consider further the way social abstraction is imagined. The culture of the simulacrum, he argues, "comes to *life* in a society where exchange-value has been generalized to the point at which the very memory of use-value is effaced."[75] We will return to Jameson's loss.

Part II

Uses and Abuses of Uses

The interventions into the discourses of contemporary art theory by Buchloh, Jameson, and Foster were intended as progressive arguments, but their emancipatory intention also had another effect. Arguments generated by radical intellectuals—as a way of resisting conservative postmodernism (with its more-or-less naked celebration of capitalism) or, in Jameson's case, with the ambition to acknowledge and explain the symptoms of a revised capitalism—generated their own political closures. It is worth looking briefly at some of the key theoretical sources drawn upon by Buchloh, Foster, and Jameson.

Readily assimilated by cultural theory, it is nevertheless revealing that Jean Baudrillard should be one of the very few theorists honored with a special insert box in the extensive survey of twentieth-century art put together by Buchloh, Foster, Krauss, and Yve-Alain Bois.[76] In his 1973 publication *The Mirror of Production*—part of the trend to question the economism of the established communist parties—Baudrillard takes on the "productivist model." But this is not a book proposing to salvage the radical heart from the flattening of official orthodoxies. In Baudrillard's view, Marx is guilty of the "fetishism of labor," leading to a false and naturalizing emphasis on utility, which downplays consumption and underestimates the role of symbolic factors. Developed on the model of the Saussurean sign, Baudrillard posits a fundamental break in the history of social need.[77] As a central component of this argument, he suggests that Marx treats use value as a "natural" and "objective" entity, as a "concrete positivity" that antecedes exchange value. Alleging (incorrectly) that Marx treats the category of use value "as distinct from any historical, social and cultural determination," Baudrillard argues that it is only "produced by the play of exchange value" and that use value is but a "code effect, the final precipitate of the law of value."[78] An edifice of gross misprisions, Baudrillard's account—which extended into his promotion of the simulacrum, hyperreality, and virtuality—thematized a discourse of social abstraction and dematerialization.

He may have recognized the inappropriateness of using simplistic oppositions, but Baudrillard still manages to transform Marx's contrast of use value and exchange value into a stand off between, respectively, "concrete" and "abstract" entities, or between "material" things and "signs." Marx, he believes, takes use value to be an "origin" and to be "natural." However, as Roman Rosdolsky points out, the contrast of use to exchange value in terms of the distinction between "natural" and "social" categories derives from Rudolf Hilferding and, following him, Paul Sweezy.[79] Indeed, Marx criticizes David Ricardo (whose conceptualization shapes how Hilferding and Sweezy approach the subject) for treating use value "exoterically," in other words, for conceiving it as a natural presupposition lying outside of the historical inflections of social exchange; in contrast—and we will come back to this point—Marx aims to understand its "role as an economic category."[80] Even as much cultural theory takes its distance from aspects of Marx's analysis (from "determinism," "presence," and, above all, from the role of living labor) while asserting "the social" emphatically (if abstractly), it unwittingly adopts the dualistic paradigm presented by Hilferding while ditching the political orientation even this model was meant to serve.

The analogy between money and language could take a very different shape to the Baudrillardian account, as it did, for example, in a 1968 essay for *Tel Quel* by Jean-Joseph Goux.[81] In order to challenge the "parallel hegemony" of commodity fetishism, and logocentrism, Goux associates Marx's "laboring trace" with Derrida's "archi-trace." Arguing that abstraction is an effacement of material signifiers which are the necessary mediating processes, Goux's account runs contrary to the one that came to prevail in cultural theory where concreteness and materiality are generally deemed to be inherently logocentric. Whatever the limitation of Goux's account, its alternative deployment of the association of sign and value form is instructive, not least for indicating the extent to which internal contestations within the linguistic interpretation of *Capital* have been forgotten. Jameson dubs the comparison of language with money and the value form, which Goux is a major proponent of, a "period comparison."[82] Associating this primarily with structuralism, Jameson also criticizes its later poststructural forms, the critiques of the problem of determinism in Marxism or of the metaphysics of presence in economic writings (the privileging of *homo economicus*).[83]

It is not only linguistic or structuralist methodology that work with homologies or analogies between cultural forms and economic forms; they are also an important component of the dialectical and antidualistic thinking associated with Western Marxism. Since the work of Lukács, Karl Korsch, and Isaac Rubin in the nineteen-twenties, the first chapter of *Capital* has been a significant point of reference for reflections on commodity fetishism and the value form. Georg Simmel's work, paralleling money and equivalence with "mental forms," is also important for this tradition of Critical Theory from Lukács, through Adorno, Marcuse, Kracauer, and Benjamin, and for considering isomorphisms between aesthetics and politics. Lukács's account of reification in *History and Class Consciousness* of 1923 extends Marx's account of commodity fetishism, charting the dialectic of quality and quantity through different aspects of social existence, and highlighting reification's "potentiation": from the experience of time, through the structures of law, to the movements of philosophical thought. Nevertheless, for Lukács, however extended and potentiated reification becomes, its apparent seamlessness is always likely to tear under the pressures of periodic economic crises, when the false totality presented by the fabric of reified forms frays to expose the alternative and contradictory totality structuring its substance.

Themes from *Kulturkritik* and Weber were present in Lukács's work, but with the Frankfurt School his moderating arguments concerning crises were abandoned. Under the "totally administered" capitalism of the twentieth century, Adorno writes, "use value perishes."[84] Similar formulations continue through his writing: "If the commodity in general combines exchange-value and use-value, then the pure use-value, whose illusion the cultural goods must preserve in completely capitalist society, must be replaced by pure exchange-value, which precisely in its capacity as exchange-value deceptively takes over the function of use-value."[85] In *Aesthetic Theory* Adorno argues that "in the age of overproduction the commodity's use value has become questionable" and that in monopoly capitalism it is exchange value, not use value, that is consumed.[86] He parallels the role of exchange value with modernism's aesthetic abstraction and art's "loss of certainty" (and likewise with "identity thinking"); in this sense, Adorno believes, abstraction "becomes a cipher of what the work is."[87] Similarly, Wolfgang Haug identifies a break in the organic link between needs and material use values, their substitution with desire and fantasy. The argument about social abstraction, told as a subsumption of use by exchange value, is extended and generalized, allied to a perceived collapse of subject and object, and to the failure of consciousness to distinguish reality and ideology. What begins as an attempt to explain societal tendencies, then, soon acquired the status of fixed fact. These theorists articulate, as one commentator puts it, "a world-historical process of reification."[88] It is an account that itself becomes reified. Yet even Adorno is not prepared to see critical value disappear completely into the rule of market exchangeability described. The "social truth" of art works, he writes, "is predicated on their fetish character"; while pointing out that the relation to art can "be modeled on the relation to actual commodity goods," he nevertheless argues that art is not reducible to its commodity status.[89] However, having blunted the tensions presented in Marx's theory of value—reflecting his social pessimism and retreat from a political reading of *Capital*—Adorno is left with the problem of identifying critical, let alone social, resistance; to rupture the link to commodity fetishism, he calls forth the residual potential of mimesis deriving from art's original magic fetishism.[90]

Guy Debord also developed his ideas from a reading of Lukács, describing the "spectacle" as the colonization of social life by the universal equivalent (money). Inaugurated by an "absolute break" in social need (note the echoes with Haug), he argues, the society of the spectacle generates a world of "limitless artificiality"; "use value has no existence outside the illusory riches of augmented survival" or the fantasies of "pseudo-use" engended by the market and mass media; the "totality of use" is replaced by the "totality of abstract representation."[91] While exchange value arises as the facilitator of the exchange of use values, it progressively acquires "autonomous power": "Starting out as the condottiere of use value," Debord notes with an enlightening metaphor, exchange value ends up waging its own war."[92] Although Debord posits a "falling rate of use value," he insists that we avoid thinking that spectacle and social reality exist in an abstract opposition in which the latter is posited as concrete and transparent; both appearance and reality, he points out, "has its share of objective reality."[93] Spectacle is not some "unreality"; on the contrary, "reality erupts within the spectacle, and the spectacle is real";[94] its "manner of being concrete, is, precisely, abstraction"; it is the "concrete manufacture of alienation."[95] On the other hand, uses and facts can be part of the false reconciliation presented to us. Debord refuses the binarisms of oppositions, such as abstract and concrete, reality and representation, reality and unreality, truth and falsehood; naming an opposition's terms should serve not to establish fixity but to delineate momentarily a field of forces within which we might trace mutations and effects. Indeed, the abstraction and rigidity of conceptual oppositions, in Debord's view, must also be understood as products of a process of reciprocal alienation. It was the task of a dialectical-critical theory to smash these versions of spectacle and to recover the "fluidity" and transience of concepts that have become frozen; being both the consciousness and the trace of movement, such thinking, he contends, displays its own "negative spirit."[96]

Spectacle is a version of commodity fetishism in which the image form is prominent ("capital accumulated to the point where it becomes image").[97] But the word "image" is slippery, as is Debord's interest, both political and poetic, in wordplay. He allows metaphoric extensions or leaps, and enables conceptual blurring, between image and representation, between "image" as literally understood and as mental representation or concept (a similar tension to the one we find in Benjamin's "dialectical image" . . . and which still echoes Hegel's discussion of image concepts). Debord insists that spectacle is "not a collection of images," but "a *social relationship* between people that is *mediated by* images."[98] The fetish elides its own production through social process, erases human and class relations and the violence to living labor.[99] His emphasis is very clearly on understanding spectacle as a condensation of social relations. As indicated by the longest, and largely ignored, chapter of *Society of the Spectacle*, Debord does not (at least not at this stage) abdicate the idea of a "unified social practice," which, he maintains, exceeds the remit of cultural negation.[100] Yet, in many commentaries on the "society of the spectacle," the term "image" is privileged and, furthermore, is treated as little more than a sign on the model of Baudrillard's simulacrum. Never mind that Baudrillard dismisses Debord's notions as passé, or that there are only tenuous parallels between them: the two accounts are frequently fused. (Jameson is just one prominent writer who has a tendency to read spectacle through a simulacral filter.[101]) In such accounts, spectacle dissolves into a process of dematerialized abstraction and spectacle is stripped of the social relations embedded in the fetish. For Marx, fetishistic appearances are not simply false, they appear as they really are as crystallizations of social relations.[102] But while Debord does not turn spectacle to simulacrum, he nevertheless emphasizes use value's illusory qualities (as does Adorno). Likewise, Debord's greater insistence on suspicion (of all appearances, facts, and realities), and his additional tendency to produce from this a social ontology of modern capitalism, allows his dialectical inversions to collapse together into equivalence's dominance-as-identity. A militant recalcitrance countervails these tendencies in his thought, but in the end Debord's reading of the categories of *Capital* seems to come from a different space. Certainly, his later readers generally adopt the tendency toward a total distrust of language and concepts—and the associated flattening of the conceptual terrain—rather more emphatically, and do so without much troubling over the question of unified social practice. This dulling of the edges of social contradiction—which accompanies, and to some extent produces, the discourses of social abstraction—is characteristic of these significant approaches in critical cultural theory, making "one-dimensionality" not only a feature of the object field of capitalism and its potentiation through subjective experience but generating that flattened condition at the level of the critical subjective outlook.

At the outset of *Capital*, Marx states that the commodity has a dual form, but, I shall argue, this does not amount to a simple dualism, still less to the types of social ontology and political prognosis often associated with it. The account now so widespread in cultural theory draws much of its force from combining Manichean clarity and metaphoric imprecision. At heart, the story of "social abstraction" is a classic tale, taking the evil-demonic form associated with modern allegory. However, the usual terms are flipped: if in classic allegories spiritual grace is lost through sin to worldly alienation, the contemporary one has us tumbling from materiality into spectralized abstraction, and increasingly what is related—Jameson being its most articulate bard—is not so much the story of commodification's onset as that of its relative densities and saturation. This contemporary version differs from most narratives patterned on the Fall: not only is there no return home to a condition prior to generalized commodity production; there is—as with Jameson's "loss of loss"—no such home at the outset. What is also striking about the narratives of the reign of equivalence is how often they further collapse their dualistic interpretation of the commodity's twofold character into a one-dimensional monism. Frequently used to provide the analytic frame for understanding the cultural forms of contemporary capitalism, this metadiscourse itself might be said to be but one of the symptoms. Like nostalgia for an imagined past of original fullness, this negative-pastoralism—which presents itself as a disabused and realistic antipastoral—may also be one that accompanies capitalism all the way to its blessed end.

The assertion that use value has been lost seems so widespread that we think we immediately grasp a profundity in Jameson's claim that we have lost even our ability to recollect it. Take a step back from his assertion, however, and the argument turns out to be less convincing. Is it really imagined that use values—or, those qualities that Jameson and others associate with them—have disappeared from memory? Have they really vanished from use? At one level, the argument is preposterous: were it to be true, we would not be fed, watered, clothed, and housed, nor, for that matter, would we be communicating through digital networks. Nor does the story make much sense as a distinction between different stages of capitalism. Not only do use values form the basis of human survival; desired or not, we are surrounded by their presence on an ever increasing scale. Far from having disappeared over the past two centuries, the production of use values has massively expanded. Without them, there would be no "society of waste." Contemporary capitalism, like any other social formation, produces objects and services for use. The alleged break within social need and the subsequent generation of "pseudo-uses," which form a central motif of recent cultural discourse, are clearly not recent phenomena, and there exists a protracted debate over wants and needs.[103] The distinction between "real" and "artificial" uses is historically defined. Today, many of us take for granted the sheer usefulness of tweezers and nail clippers but Adam Smith saw them as "trinkets of frivolous utility."[104] Equally interesting is Smith's additional observation that what fuels desire for such devices is not their usefulness but rather "the aptness" of the contrivances "fitted to promote" them: the invention of "new pockets" in which to keep the "toys." Similar discussions can be found throughout the texts of early political economy, such as in Nicholas Barbon's: "The greatest number (of things)" serve not to supply the wants of the body but those "of the mind," and the "desire" produced by the mind is as important as the "hunger" produced by the stomach, his argument goes.[105] Advanced during the recoinage crisis, his "state theory of money," which insists that the value of coins be decreed by law (rather than based on their metallic content), presages the break that for many recent cultural theorists occurs only with the end of the gold standard. The essential outlines of today's cultural discussion, then, can be found already in texts preceding universal commodity production. If modern semioticians are inclined to see value as the effect of arbitrary convention, Barbon's emphasis on convention differs in being "subjective," with value swinging between the whim of royal decree and the unpredictability of speculation. According to Barbon and Smith, even if the augmentation of survival involves trivial "pseudo-uses," we should not understand it to be a bad thing.[106] And, of course, most augmentation is, in reality, far from trivial.

Struggles for a society based on social needs rather than private profit form crucial axes for those committed to social emancipation. In the light of environmental sustainability and global inequalities, there is a serious debate to be had over future possibilities and limits of augmentation, but the discussions about how to weigh potential with social and environmental costs, or how social wealth might be distributed and developed, should not be confused with cultural theory's arguments over the generation of pseudo-needs and the manipulations of desire. Resisting dispossession in defense of shanty districts in India's cities or challenging the repossession of houses whose mortgagees have defaulted; fights to save social facilities and services, or for access to the latest anti-cancer treatments or retroviral drugs; efforts to defend jobs or prevent wage cuts: all are fights over the value form itself, for the assertion of living over dead labor, for the prioritization of social use over capital's pursuit of greater profitability. In this sense, use value is a political category and the focus of many counter-campaigns to capital. However, in challenging cultural theory's narrative of the total domination of exchange value (an account claiming emancipatory critical power), it would be wrongheaded to attempt to construe use value the begetter of a rival discursive dynasty. Rendering capitalist social relations in such terms falls into the same trap, impoverishing our understanding of social relations by reducing them to the shorthand of highly condensed metaphors. The romanticization of societies before universal commodity production—the pastoral conception of use value—is neither coextensive with Marx's account nor even with all Marxisms. (István Mészáros, for instance, has noted the extent to which historical societies dedicated solely to the production of uses were often unable to fulfill even basic needs.[107]) As Marx himself notes, modern yearnings for past social formations should be understood as the products of capitalism. What will be important to recover, then, are the relations between use and exchange value for the understanding of the value form itself, and to do so from writers who, while advocating a dialectical approach, present us with a dialectic that withers. Doing so necessarily involves reemphasizing the place of use value in this account; however, the substantial point rests not on a concomitant downplaying of the value form's abstractions or of the importance of "real abstraction" but on the political consequences of a discourse that articulates use value's all-but-total diminution. There are more than just formal issues of textual interpretation at stake here; invariably, accounts of the value form claim social validity and disseminate prognoses on social emancipation. The story of social abstraction is, at best, a loose metaphor for describing the tendencies of social relations under capitalism. Were this metaphor to contribute to the mobilization social resistance to capital's logics, one might be more sympathetic, but it has become more than a casual abbreviation, and now props up a fatalism based on a conception of an undifferentiated totality before which we are powerless.

Determinate Social Form

It is easy to lose sight of the fact that Marx himself never refers explicitly to "social abstraction" in the way that has become commonly assumed in the theoretical discourses of contemporary art and culture, where the term attains the status of a distinctive and fetishized process. Despite having been repeatedly attributed to the first chapter of *Capital*, the story that use value gives way to exchange value in a process of social abstraction is not one that is told. *Capital*—and chapter 1 above all—has often stymied its readers. Marx acknowledges that his opening analysis of the commodity form, where he aims to conceptualize the very heart of capitalist social relations, presents difficulties. The challenge turns on two related issues: his discussion of the "form of value" and the technique of analysis he uses (that is, his historical-logical method and his deployment of "the force of abstraction").[108] The discussion of the form of value now occupies section 3 of the first chapter, but it has an interesting history: starting out as an appendix (which Marx only included in the first edition of *Das Kapital* at the final hour), it was then revised and integrated into the main text for subsequent editions.[109] Marx reports how he came to realize that his analysis of the value form would be "decisive for the whole book."[110] Far from charting the disappearance of use value, Marx in this section outlines its distinctive relation to exchange value.

The analysis of the form of value is conducted in four stages: from the "Elementary, simple or accidental form of value" (A), through "The Expanded Form of Value" (B), to (C and D) "The General" and "Universal" forms. We progress from the chapter's original scene of commodity exchange where twenty yards of linen is exchanged for one coat, to the emergence of money as the universal medium of exchange. Echoing approaches taken in the classic texts of political economy, the first stage is often equated with an earlier point in the development of commodity production where a social unit (a family or tribe) exchanges its surplus only as an occasional or a peripheral practice. But the staging is not straightforwardly part of an imagined historical chronology. The encounter between the weaver of linen and tailor of coats need not be understood as describing the premarket economy, as often presumed, and the exchange might just as readily be taking place between two capitalist merchants. Indeed, another temporality is at work in Marx's analysis and a logical priority is also at stake. Marx warns us that his is a retrospective analysis in which the method of inquiry (proceeding from the complexity of capitalism's economic system to its simple elements) runs in the opposite direction to his method of presentation.[111] (And he plays with the language of simplicity and complexity: the complex manifestations of value, which we more readily encounter in the world around us, are often the easiest for us to recognize, whereas we struggle—as we do with chapter 1—to grasp the more abstract simple forms.) The first stage, the elementary form of value, is as much analytically excavated as it is historically derived from the money form. As Marx says, it is the "almost imperceptible outline" detectable in the elementary form of value that he is after and which, he argues, lies in the commodity's "dual nature": its being both an object of utility and an object of exchange. The commodity's double character is expressed as use value and exchange value; it corresponds to the twofold character of labor: the concrete labor of specific human activities and labor taken in the abstract.[112] Against the "bodily form" or "physical" form, or the "coarse materiality" of use value, Marx notes, reiterating his discussions in sections 1 and 2, exchange values have "a purely social reality," that is, they are the "expressions or embodiments," as he puts it, "of one identical social substance, viz., human labour."[113]

"Form of value" is used in an additional sense: to refer to the "two poles of expression" of value, the *relative form of value* and the *equivalent form of value*, which manifest within the process of commodity exchange. The analysis of these two expressions of value runs through the four stages (A–D) of section 3.[114] At one level, this distinction between the poles of expression is straightforward. In the elementary stage, where "20 yards of linen are worth one coat," it is the linen that is said to be in the relative form (since it is the commodity whose value we want to express), while the equivalent form is taken by the commodity *in which* we are seeking to express the linen's value (the coat). In the elementary form, these roles are arbitrary and can be readily reversed—"one coat = twenty yards of linen"—and different commodities may be substituted in varying proportions; only with the universal form does the commodity adopting the equivalent form become stabilized on the money commodity. At another level, the distinction is more difficult to grasp. The expression of value, Marx repeatedly notes, is at the root of the mystery of universal equivalence (money) and its power as social form.[115] It is locating the relationship of relative and equivalent forms of value that Marx has in mind when he states that the secret of commodity fetishism's enigmatic nature lies in the form of value. The antagonistic interplay between these two poles of value expression dominates the discussion in section 3, comprising a full one half of the opening chapter. Attention to these passages reveals that the heart of the difficulty of reading this work lies in grasping the dialectic in which the relative and equivalent forms of value exist in a relation of mutual opposition *and* mutual dependency.[116]

Until one commodity enters into exchange with another, a commodity's dual nature remains latent; it is the process through which "the social relation of commodity to commodity" is formed that enables the external presentation of this twofold character. "The opposition or contrast existing internally in each commodity between use-value and value," Marx notes, "is, therefore, made evident externally."[117] As these two characteristics externalize, they become attributed to the relative and equivalent commodities. The commodity in the relative form "figures directly as a mere use-value" (its "bodily form . . . figures only as a use-value"), while that serving as equivalent "figures directly as mere exchange-value" (its bodily form figuring "only as the form or aspect of value").[118] In the *Grundrisse*, the relation between relative and equivalent forms of value is more formally outlined in openly dialectical language. In terms both spatial and temporal, Marx describes a process of internal splitting and external separation.[119] The commodity's "double, *differentiated* existence," he argues, "must develop into a *difference*, and the difference into an *antithesis* and *contradiction*."[120] Here and in *Capital*, Marx traces how latency becomes manifest, how social form is reduced to one-sidedness, how a fluid state is rendered static. We encounter duality in a number of ways: as latency, as differentiation, as difference, and as antithesis. With its double characteristics now autonomized from one another, this final manifestation—the result—induces the conditions for Marx's discussion of commodity fetishism in *Capital*; in other words, what Marx sketches in the *Grundrisse* is the process of the fetish's emergence.

Considering the heart of capitalist social relations in terms of the commodity's latent duality however, is not to say that we should therefore produce a dualistic schema (let alone that we should then further translate this into a passage "from . . . to . . ."). Even for the equivalent commodity, the one that "*figures* directly as mere exchange-value," use value has far from evaporated; on the contrary, its "bodily form" is the bearer [*Träger*] of exchange value. Money is the commodity whose distinctive use value is to serve as universal equivalent for all other commodities. But the universal equivalent is additionally distinctive because its use value no longer functions as content (to this extent alone use value could be said to "disappear"). However, it plays a crucial role as form, as Marx insists in his retort to Adolf Wagner. Reviewing *Das Kapital*, Wagner complains that Marx overlooks the relevance of use values. On the contrary, Marx clarifies, "for me use-value plays an important part quite different from its part in economics hitherto, but *note bene* it still only comes under consideration when such a consideration stems from the analysis with regard to economic formations, not from arguing hither and thither about the concepts or words 'use-value' and 'value.'"[121] Marx takes economic forms to be the clearest indices of the historical changes to the organization of social relations. His concern is to establish where and how categories often assumed as transhistorical specifically operated within—or were determinate for—generalized commodity production. The historical specificity of exchange value's dominance is widely accepted today, but use value's disappearance is often taken, mistakenly, to be the corollary. Looked at from the perspective of simple commodity exchange (the topic of chapter 1), the qualitative heterogeneity of commodity uses (the content of use value) is of little relevance for understanding the development of the value form, and to this extent the category is presupposed (that is, a presupposition in the strong sense, not a mere assumption that posits use value as natural).[122] However, when considered from the perspective of form—more precisely, when considered as a historically determinate social form—use value becomes critically important in the case of two highly significant commodities: the money commodity and labor power. We have already addressed the use value of money, but labor power, or labor as commodity, also has a specific use value: the capacity to create value. By extracting surplus value from labor—that is, the difference between the value created and that received by the laborer in exchange for expending his or her efforts—capital is able to increase its own exchange value. As a commodity, labor power is thus seen by Marx as central to capitalism's social formation. These two commodities—money and labor power—are distinctive because their constitutions take us the very heart of capital's social mystery; the role of use value as determinate social form is decisive. These exceptional examples are presented in the opening chapters of *Capital.* Marx then extends his analysis further, taking into consideration more developed expressions of value, and more concrete examples of reproductive processes; as he does so, the role of use value *as a specific social form* becomes even more important.[123]

Understanding use value as a socially determinate form is central not only to Marx's account of the historical specificity of social forms and categories but also to his analysis of exploitation and social contradiction. Attending to the specific social form of use value—as opposed to focusing on use value as content—alters the perspective on the cultural account of social abstraction. Cultural theory generally turns the four stages of value into a historical narrative of the disappearance of use value (this theory's fetish of abstraction not extending to a recognition of methodological abstraction, still less to one understood dialectically); it sidesteps the discussion of relative and equivalent forms, reducing capital's distinguishing form to exchange value alone, elevating (and formally abstracting) the discussion of "real abstraction"; dispensing with a labor theory of value, it dislodges human labor from any role aside from being an abstract social nonsubstance which enables its challenge to use value's alleged natural constitution.

Jameson's Balloons

One of the prime rehearsals of the account of social abstraction can be found in Jameson's writings. As indicated by his continued interest in utopia, science fiction, cognitive mapping, and archaeologies of the future, an impulse to figure a world beyond capitalism has remained central to his work.[124] Jameson also constantly returns to the problem of the "reality of appearances" and he seeks to remain close to Marx's account of "commodity fetishism" in which the very objectivity of a social form harbors a contradiction. As indicated in his earlier work—by his programmatic treatise "Towards Dialectical Criticism" in *Marxism and Form* or by his contribution to *Aesthetics and Politics*—Jameson has a sophisticated understanding of dialectics, of inversions and contradictions, the interplay of necessity and freedom, and the interaction of concrete and abstract thinking.[125] Moreover, Jameson is generally wary of absolute differences, where, as he puts it in characteristically Hegelian language, "the materials of the text . . . fall apart into random and inert passivity, into a set of elements which entertain purely external separations from one another."[126] Yet, for all the evidence of Jameson's capacity for dialectical thought at the level of the particular textual reading, the net effect of his argument leans toward a one-dimensional account of social abstraction. Indeed, the strengths of the former often mask the weaknesses of the latter, and a rhetorical strategy of "I know but nevertheless" is a regular feature of his *modus operandi*. As we will see, Jameson seems unable to think the value form, especially in its most recent manifestations, as anything other than the type of simple abstraction premised on the linguistic analysis. He himself criticizes the understanding the rise of the money form as "the supersession of concrete use and function by an essentially idealistic and abstract 'fetishism of commodities,'" attributing this position to "standard Marxian economics"; yet he goes on to replicate this very problem, the main distinction being that his efforts are directed (unsuccessfully) at making that abstraction less ideal.[127] He also consistently challenges the structuralist model itself, including the weakness of its homologies with the value form; but again, we find echoes of this very model in his own essays on late-capitalist culture, although he seeks to replace a weak homology with a stronger variant.[128] Returning to Jameson's point about memory: we may have lost the experiential memory of life in a society where the value form was absent or only partially developed, but this is an altogether different statement from his assertion that we can no longer summon even the memory of use value. What might reasonably be described as a tendential shift (in which social relations are subject to the generalization and intensification of the role of equivalence), or be a claim that is primarily intended to have rhetorical force, is turned into a rather tendentious social ontology based on the "logic of the signs." Already characteristic of his essays on postmodernism from the eighties, the problem continues explicitly in writing from the late nineties where Jameson considers the relation between architectural configurations and finance capital.[129]

In "The Brick and the Balloon: Architecture, Idealism and Land Speculation," Jameson's first concern is methodological: to argue against reductive conceptions of the relationship between culture and economics. Jameson's immediate target is Robert Fitch, whose book *The Assassination of New York* exposes the role of land speculation. Criticizing what he sees as Fitch's "conspiratorial" conception of history, Jameson proposes that we understand land speculation in terms of capital's structural logic. (In fact, there is a subplot: Fitch distrusts theorists who explain capitalism through "logics"—that is, theorists like Jameson—a type of argument that, in Fitch's view, downplays the role of individual speculators and, more importantly, marginalizes the place of agency in resistance to them.) Challenging the reductiveness of the conspiratorial approach, Jameson calls for mediation between the aesthetic and economic dimensions of late capitalism. He wants us to maintain a dialectical sensitivity to the relative autonomy of these dimensions while simultaneously grasping their place in the social totality. Mediation, Jameson insists, is a "two-way street," but, as I will argue, he does not go on to prove his case for mediation by example. (Nor for that matter does Jameson respond in any way to Fitch's charge: at no point, even in his own terms, does he address the question of material or political agency in either its pro- or anticapitalist forms.)

Continuing the project to explore the relation of aesthetics and politics, Jameson's considers how cultural and aesthetic forms are (mediated) representatives of universal equivalence and money. If Siegfried Kracauer's "mass ornament," say, posited links with Fordist and Taylorist systems in the interwar period, today's task is to probe those with finance and fictitious capital. I have alluded already to Jameson's criticisms of homologies, especially the limitations of their structuralist and New Historicist lineages: how they allow imprecision which nevertheless yokes together distinct phenomena into identities, how despite claiming to avoid ontological priorities their connections are often "as crude and vulgar" as the most narrowly determinist Marxism.[130] But his work on postmodernist culture, and the "Brick and Balloon" in particular, might be seen as an attempt to bring together the explorations of homology by structuralist writers, New Historicism, and the dialecticians of German Critical Theory, to address mediation, structural parallels, and immanence. (It may be not without significance that Jameson describes homology as a method that "disengages an abstract structure."[131]) When Jameson states that finance capital is one of late capitalism's distinctive features, he is not wrong, and he is certainly aware that it has a long history. Nevertheless, this observation is marshaled to perform a specific function: it reinforces his separation of postmodern from modern periods.[132] Taking some of the more general characteristics of the credit system (few of which are, per se, unique to capitalism, let alone to the most recent period of capitalism), Jameson proceeds to extend some broad cultural homologies. It is the pattern and content of this operation to which we should attend.[133]

Late-capitalist culture, Jameson argues, produces heightened forms of abstraction, characterized by transformations in its temporal and spatial logics that are patterned on the effects of the market in futures.[134] As Jameson sees it, "time itself" has been reorganized on the model of financial speculation, analogous to the anticipation that a successful financial gamble could pay off. He asks, then, that we probe how the patterns of finance capital register in aesthetic forms and sensibility. The "new relationship to the future," he argues, "is now the final link in the chain which leads from finance capital through land speculation to aesthetics and cultural production itself . . . to architecture."[135] The mediating term that Jameson proposes is Baudrillard's simulacrum ("surely the most inventive exploration of the paradoxes and afterimages of this new dimension of things"), which he describes as an "abstraction from beyond the already abstracted image."[136] This "abstraction beyond abstraction" describes not only the simulacrum; it is also Jameson's characterization—evocative, vague (and inaccurate)—of finance capital.

Although he has a longstanding interest in architecture and space, Jameson's work largely pursues a politics of time, in narrative, but most notably through the temporality of the dialectic itself. In "The Brick and the Balloon," he responds to critical geographers such as David Harvey (who argue that Marxist theory has overemphasized history and temporality). By way of the anticipatory effects of fictitious capital, Jameson tries to integrate the refiguring of time with an analysis of spatial form.[137] What is the nature of his "final link" in this ambition to bring together a

Marxist cultural analysis of architectural space and the value form? We might perhaps expect a consideration of how the "structural reorganization of time" (or the temporality of delayed gratification being located with financial speculation) relates to volumetric organization in architecture. However, this is not what Jameson provides, and again we are returned to an embarrassingly narrow conception of abstraction. Drawing on Charles Jencks's description of recent architecture (Jencks's writings can be found underpinning Jameson's work on postmodernism), Jameson describes the emergence of abstract spaces, where superficial skins enclose "extreme isometric space." With "energetic distortions," these spaces, it is argued, exaggerate the free plan promoted by the modern movement.[138] Late-capitalist isometricity, however, strips form of any relation to distinctive business activities and—in an argument analogous to those that focus on the common platform provided by digitalization—its spaces accommodate interchangeable functions. The waning of affect and historicity, which Jameson discusses in his classic essays on postmodernism, are here supplemented by a waning of function—or, rather, function is now achieved through the waning of concrete specificity. This transition in building type is turned into a metaphor by Jameson, following Jencks, as "from brick to balloon": from a type of architecture that emphasizes weight, mass, and density to one determined by contour and volume. Once again, what is being suggested is a shift from material to immaterial qualities. Nodding toward not only Marx on the value form but also to Rem Koolhaas's manifesto, *Delirious New York*, Jameson has isometric space represent "the very element of delirious equivalence itself."[139]

There are echoes here of the traditional distinction of construction methods and associated symbolism made in architectural history between Romanesque and Gothic cathedrals: massive load-bearing stonework replaced by flying buttresses and soaring ribbed vaults; "heaviness" replaced by "lightness"; dark solemnity supplanted by spiritual light. We would be fools, however, to imagine that cathedral buildings are any less load bearing; Gothic structures do not negate weight, they alter how it is structurally transferred. The "brick" functions rhetorically for Jameson as the dark and weighty ground against which newer isometric forms are contrasted, but the metaphors are limited. Bricks can act as "balloons" but be no less "bricks"; they can provide infill for curtain walls or for traditional "balloon frames." Bricks—to push the (literal *and* metaphoric) problem further—are, and have long been, among the most adaptable of building components; modularity is their forte. We might even say that "equivalence" is in their nature (and were we to pursue *this* metaphoric route, we could produce a very different "cultural logic" to the one set out by Jameson). I am pushing Jameson's analogy too far, of course, taking his metaphor too literally; but one has only to prod a little to see its curtain walls collapse—and not just the folding of the metaphoric vehicle but also of the social homology it shores up. And, in the end, Jameson himself literalizes the metaphor.

Jameson acknowledges the analogy between isometric "balloons" and Gothic architecture, pointing us to what he sees as the limitations of the comparison: today's isometric space may be "light," he notes, but it does not symbolize the religious uplift that was associated with the cathedrals at Chartres or St. Denis; instead, it "dematerializes without signifying in any traditional way spirituality."[140] As he pursues this point, the problems at the core of Jameson's argument become clearer: "where the free plan posited an older bourgeois space to be cancelled, the infinite new isometric kind cancels nothing, but simply develops under its own momentum like a new dimension."[141] Isometric space no longer has any point of resistance: it is not a determinate negation, scarcely acceding even to the status of an indeterminate one; rather, it is more an infinitely spreading *informe* existing without historical coordinates. Jameson's argument is most explicit where he turns to discuss how this situation relates to the characteristics of finance capital. Embodiment and weight become attenuated and abstracted in isometric architecture just as financial capital is "abstract" or "materialist sublimation."[142] In "Culture and Finance Capital," he associates the money form with modernism, describing it as semiautonomous: "both abstract (making everything equivalent) and empty and uninteresting, since its interest lies outside itself . . . it directs attention elsewhere, beyond itself, towards what is supposed to complete (and also abolish) it, namely production and value."[143] Finance capital, however, achieves "full autonomy"—a "dimension in its own right"—being "a play of monetary entities which needs neither production (as capital does) nor consumption (as money does): which supremely . . . can live on its own internal metabolism and circulate without reference to an older type of content."[144] Another rehearsal of his modern-postmodern distinction—where the abstraction of modernism (aesthetic and social) is supplanted by the hyperabstraction of late capitalism—Jameson here also rehearses a rather dubious understanding of financial capital.

Except at the most superficial level of interpretation, a fully autonomous "play of monetary entities which needs neither production . . . nor consumption," which lives off "its own internal metabolism," is very precisely not what finance capital brings into being. It is Harvey's discussion of fictitious capital, and its role in the land markets, to which Jameson appeals, and which is central to providing "the final link in the chain" between the isometric architectural space and the value form of financial capital. However, what Jameson chooses to draw out from Harvey's discussion—and what he ignores—is revealing. Indeed, the elision of the central aspects of Harvey's account, on the one hand, is matched, on the other, by Jameson's hypostatization of those features upon which he does choose to alight. Jameson here seems subject to the very fetish that both Harvey and Marx criticize: the belief in capital as self-generating source of value (long characteristic of capital's very own utopias and dreamworlds) and the absolute autonomy of financial and fictitious capital.

In its underdeveloped nineteenth-century form, the credit system already represented the most prominent, and everyday, example of fetishization for Marx: "the fountainhead of all manner of insane forms," where, among other things, the accumulation of debt was mistaken for the accumulation of capital.[145] In his discussion of the role of the credit system and interest-bearing capital in the third volume of *Capital*, Marx demonstrates how credit was used to overcome the barriers presented to the circulation and realization of capital: to iron out the delays to the realization of capital between production and exchange; to moderate the inflexibility and immobility of fixed capital; and to coordinate between (or centralize) the many individual forms of capital and the interests of capital in general. The credit system forced capital's intensive and extensive expansion, and spurred its internal transformations, the growth of a world market and geopolitical reorganization. Fictitious capital emerged as a type of interest-bearing capital that, instead of being guaranteed by firm collateral, speculates on a "marketable claim upon some future revenue," that is, the revenue from the anticipated exploitation of labor and appropriation.[146] It has long since been institutionalized within capital's day-to-day existence; the intense financialization of the economy under neoliberalism being the most recent manifestation and intensification of this institutionalization. Its shadow banking system and trading in credit derivatives, which came to such spectacular crisis in 2008, may have been a result of deregulation (to this extent "autonomous," although even neoliberal deregulation was never as absolute as was ideologically proclaimed). However, the hyperventilating financial economy was not autonomous from the "real economy"; mutually imbricated, the former's exponential growth through the nineteen-nineties was an important expression of the latter's diminishing returns for capital investment.

There are a number of key points that Harvey, for one, makes about the credit system in general and fictitious capital in particular, each of which calls into question Jameson's reading and conclusions. Firstly, Jameson bypasses the contradictions inherent in fictitious capital, ignoring its double-edged character, and deploying Marx's and Harvey's accounts selectively. It may remove the barriers to the circulation of capital but, as Marx puts it, credit "suspends the barriers to the realization of capital only by raising them to their most general form."[147] The "energetic distortions" produced by, or "irrational exuberance" resulting in, the overaccumulation and overinflated value of fictitious capital turn into crises, at which point investors scramble to realize the value of their "assets" before they evaporate in the rush to relinquish rapidly tumbling share values. (Following the collapse of the country's banks in Autumn 2008, some Icelanders reached for the "solidity" and international exchangeability of Rolex watches; and mindful of the history of currency devaluations, the German middle-classes were reported to be turning their Euro to gold at such a rate that the market ran out of supplies.) Through these cycles of credit-based expansion, speculative fever, crash, and recovery, the credit system accelerates the development of productive forces by suspending the barriers to realization of capital; but it also accelerates crisis formation (by delaying and thereby exacerbating capital's problems).[148] Harvey refuses to reduce finance capital to an abstraction—let alone an abstraction beyond abstraction—and constantly reiterates fictitious capital's material bases.[149]

He also adds a further, somewhat bolder (and contentious) argument: drawing on Marx's discussions, and his own analysis of the New York financial crisis of the seventies, Harvey sees fictitious capital not just as speculative froth on the surface of real production, and not only as playing the important processual role for capital outlined above, but as latent in—and necessary to—capital's very concept.[150] "Necessity" is a figure that often features in Jameson's writing, not least as "logic," but here he adapts—and transforms—Harvey's argument, positing finance capital as the domain of "necessary fictions." It may well be so (traders have to convince themselves that they will not be the ones to be caught in the crash, and must imagine that they will be smart enough to have jumped from the fictitious ship before it sinks). However, the figure of "necessary fiction"—especially the word "necessary"—allows Jameson to invoke an antiidealist and nonfetishistic response to finance capital while actually delivering the opposite. It is a strategy that can be compared to those linguistic interpretations of value that rhetorically summon a link to (social) materiality by asserting real abstraction, yet do so while erasing social contradiction from the field and by transforming real abstraction into the reality of a one-dimensional abstraction. Jameson argues against the linguistic model of the New Economic Criticism and New Historicism, but his essays on finance capital bear the marks of the same thought process: emphasizing financial capital as a fully autonomous form with no connection to production or consumption, he posits an absolute break. For linguistic accounts of the value form, the break that predicates the unleashing of value is located variously, the most favored being the emergence of money as token of value and the abolition of the gold standard. Before the rupture, so such accounts go, value was grounded in, tied to, or guaranteed by actual commodities, this being no longer the case after. For some, this schema is mapped onto the abstracting of exchange value (after) from the materiality of use value (before). Jameson pushes the point further still, with the relative autonomy of money giving way to the absolute autonomy of financial capital.

Those who see the arrival of metallic tokens, paper money, plastic cards, or credit, or the end of the gold standard as the autonomizing of value from materiality essentially read, or revise, Marx with a limited structuralist model. It is not in this sense that use value or materiality (or their alleged demise) could be said to be at stake. Understanding the issue of use value from the perspective of socially determinate form, it is less a matter of whether money is an object that is composed of precious metals, underpinned by guarantee, or circulates as cheap tokens, as credit or as the electro-digital pulses of today's international trading. Developments in the history of money are important (and are certainly far more complex than suggested in the standard cultural narratives), but chart neither the dissolution of production of use values nor that of social contradiction that is at the heart of Marx's account of the value form.[151] Cultural theory both overestimates and underestimates the effect of the end of the gold standard. It overstates, for example, the moment as some fundamental historical rupture in capitalism, after which financial sign forms float autonomously, ignoring the degree to which, as Marx outlines already in the *Grundrisse*, the commodity's internal contradiction already predicates the value form's autonomization. At the same time, the significance of the generalized move to inconvertible or fiduciary money is underestimated: the increased destabilization and entrenching of the value form's problems. Marx himself abandons the treatment of money as symbol and the analogy of money and language that he still employs in the *Grundrisse*, thereafter emphasizing instead his central argument that money should be understood as a social relation. In the context of capitalism's social relations—indirect relations between people and direct relations between their products—the value form acquires a status of independence, becoming in the process the object of commodity fetishism. Marx's accusation of fetishism is directed primarily at economic theorists who think value emanates naturally or magically from the value form itself, rather than recognizing that the form is produced by human activity. The *Grundrisse* itself already contains the critique of such perspectives. One of Marx's target's here is not a mainstream economist but Alfred Darimon, a Proudhonist. Darimon hoped to displace the social sovereignty of gold and silver, and with them the moral evils associated with money, with a system of time-chits. Marx's argument here turns not on whether money has specific physical or chemical properties, nor on its role as sign, but rather on the problem that the time-chits system could not sidestep the social relations in which they were embedded.[152] Indeed, elsewhere in the volume Marx even mentions the extinguishing of "the memory of use value" in the "incarnation of pure exchange value." Far from upholding the type of the generalized claim made by Jameson, the context is a discussion about how, while the Incas and Aztecs used gold for making jewellery, nineteenth-century economists assume gold to be inherently money—that is, the context is a critique of how these economists misread, and fetishize, the value form.[153] The central point is that the "incarnation of pure exchange value" *is* the moderns' fetish.

If the nature of world money has mutated since the removal of the dollar gold standard, from a condition of (relative) stability to one of the instability of fiducial relations, this does not mean that it is no longer related to social exploitation or divorced from production and consumption. The prominence of financial and fictitious capital reorients the shape and the temporality of these ties. Instead of resting values on, or wresting them from, existing commodity production and exploitation, it gambles on anticipated profits and the extraction of surplus value. Such profits cannot be guaranteed, of course—but, then, neither could those of the conventional production-originated model (itself a methodological abstraction that is too often conflated with reality); would the commodities sell at, or above, or below their value? Or would they languish in the warehouse, unable even to find a buyer? There was no surety unless contracts could be set up to guarantee sales in advance, the basis of today's futures, options, and swaps—all attempts to hedge such uncertainty—which are described in volume 3 and which are embedded, Harvey suggests, in capital's very premise. Even here we would want to be more circumspect about drawing homologies from such temporal reorientation to the future. If we take the previous points seriously, the reorientation is never just from past (prior values) to future (anticipated values), but rather a complex formation and integration of both features simultaneously.

The isolated cycle of production-exchange-consumption is an abstraction from the actual working circuits of capital for the purpose of analysis; the future-orientated model that attempts to displace it is no less so. The latter, however, seems not to recognize the methodological force of abstraction at work in the former; and in its own case, only identifies abstraction as an external social process. If (methodological) abstractions are to help us grasp and analyze the features of capital, and do so better than direct descriptions of reality, we need to exercise greater caution with their categories. Even taking Jameson's argument on its own terms, and even were we to describe the temporal and spatial characteristics elicited by today's financial markets, more nuance would surely be demanded. Let's just take the question of time. Projections into, and gambles on, the future depict but one aspect. (And just how new are such temporal anticipations anyhow?) If we really want to construct this sort of argument, why not focus on, say, the tendencies to abolish time—or, at least, reduce it to seconds—in hyper short-term investments of currency trading facilitated by computerization, all those efforts to make a profit on miniscule differentials between currencies or across the money markets? Equally, other features emphasize retrospective temporalities, slow movements, or the stretching of time. And we have yet to try to translate any of these different temporalities into their spatial or architectural expressions. It seems, then, that making homologies—whether simple or complex ones—tends to be a particularly selective activity, and it is worth noting the type of features that are typically privileged in cultural theory: abstraction (never concreteness), amateriality (rarely materiality), futurity (less so past or surviving forms), speed and fragmentation, or, with some notable exceptions, capital's autonomy (not labor's). Such theory likes a simple story composed of complex-sounding terms, but avoids confronting actual complexities and contradictions (to which Foster, in his criticisms of Jameson, is partly alert). Significantly, this approach refuses to confront the question of how contemporary credit and debt (further manifestations of the value form) embed social relations. The narrative suppression of use value and the simultaneous cutting free of finance from the production of surplus value have allowed many writers to lose sight of a central element in the account of reification and commodity fetishism. The eruption of periodic crises and the social contradictions they bring into focus remain outside their account. The consequences of the argument are to eternalize capital, to endow it with the smoothness of a well-oiled machine, to use the Fordist metaphor, or, to use its successor, the uninterruptability of digital sequences of zeros and ones.

While nodding toward Marx's analysis of the fetishism of commodities, such explanations retain little basis for understanding the fetishistic process; worse still, they proceed to offer up the fetish itself in new guise. By rendering the commodity's twofold nature a dualism at the social core (rather than a contradiction), by simplifying a narrative of social abstraction, and by evacuating social exploitation from its accounts, cultural theory traverses the account of commodity fetishism only to return us to a fetishistic social theory and an associated fatalistic (or, occasionally, voluntaristic) politics. Cultural theory, in its reading of Marx, not only assumes a sense of abstraction loosely inspired by Weber, it also crudely literalizes Marx's own deployment of abstraction, making from the two a generalized cultural metaphor. As can be seen in Jameson's essays on financial capital, what is turned into a metaphor and then literalized at the level of social theory is then turned into a metaphor again in the analysis of cultural and aesthetic forms. Paradoxically, in the name of an antireductive and dialectically mediated Marxism that, refusing any taint of nostalgia, demands we address capitalism in all its contemporaneity and presentness, what has been entrenched is a reductive social theory, a simplistic homology, and a negative pastoral. Instead of basking in the radiant splendor of the old fetishes, we are now left to gape queasily before their new formlessness.

Afterword

Abstract and Transitive Possibilities

The Moscow rush hour. A collective of Russian artists linger at a busy street crossing in Barrikadnaya, an area in the city's northwest. Cars and lorries are in motion; artists and commuting workers wait. At the last moment—as the lights change, signaling the traffic to halt and the assembled pedestrians to proceed—the Radek Community unfurl their red banner "Another world is possible" and accompany the throng across the junction (see figure 5.1).[1] From the allegory of the crossroads to the confrontations of traffic and people, stasis and movement, possibility and impossibility, the group's deceptively simple action does not merely question the nine-to-five mentality but unleashes critical reflection on aesthetics and politics at the turn of the century. The performance is resonant, of course, with the photographs and film footage documenting demonstrating workers, soldiers, and peasants—and also with the history of orchestrated crowds. From the restaging of the 1917 Revolution for anniversary celebrations or for Sergei Eisenstein's films, to the parades of the Five-Year Plan photographed by Alexandr Rodchenko, these are masses made for the lens, sometimes deliberately created by artists (Radek's being something of a Surrealist *object trouvé* or Duchampian readymade; the video's deadpan narration spoken and their slogans written in English, gearing their video recording for screening at international venues). We are prompted to think about the differences between crowd, mass, assembly, and multitude; between populace, people, and publics; the distinction between aggregates of individuals and collectivities; the history of Marx's ideas on the self-

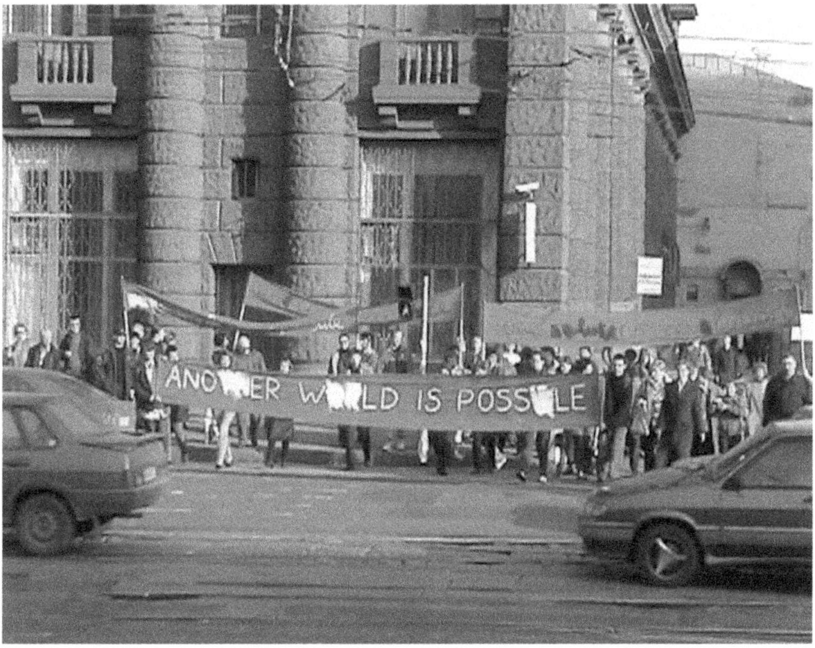

5.1 Radek Community, *Manifestation*, 2001, video still of action in Barrikadnaya, Moscow. Courtesy of Petr Bystrov on behalf of the Radek Community.

knowledge of the revolutionary class. The very notion of avant-gardism—in which this action participates and on which it comments—is brought up for consideration; and, by way of both the work's geographical setting and metaphoric association, so is that of vanguardism as an organizational model and its intersection with the history of councilist practices; or, by way of the emblematic force of this conjuncture of artistic gesture and political action, of the limits of substitutionalism—here of art for politics just as much as the will of a small group for mass social action. Human figures are there for a brief moment, but, taking their multiple directions, they quickly evaporate. Emancipatory expectations are invoked, underscoring all the more acutely the bitterness of dispersion, the crossing's command to wait being felt more potently as the pedestrians actually traverse the street. Fissures open up between what is directly proposed to us by the performance (the chance of a better world) and what it contains (a mass of individuals largely indifferent to, if not totally unaware of, the action of which they are part); or between the proposition and its situation (Russia, a decade or so after the collapse of Soviet communism and in the grip of the commodity's colonization of everyday life, or, broadening out our perspective, the context of capital's rapid global expansion at the end of the twentieth century); or between "art" and "life praxis," autonomy being the condition of the artistic action and its critical space (its freedom) but at the cost of social irrelevance and impotence (its unfreedom). The vacuity of the gesture, however, is not entirely what it seems: more purposeful and angry than nostalgic or poignant, the work exceeds the irony of its "post-punk counter-Dada" status to signal frustration with its own condition of "emptiness." Forcing temporal collisions in and with the present—whether by history itself, or by the future orientation of the anticapitalist movement, or even by what were once hopes for the future (whether of 1917 or Perestroika)—the work condenses concerns pertaining to the aesthetics and politics of social possibility. The Radek Community's action can be taken as an exemplary instance of the subject to be addressed in this afterword: the politics of emancipatory anticipation in both art's practice and theory.

The majority of the historians and theorists of art considered in *Dialectical Passions* find their projects situated within both the strictures and openings provided by this same problem. Their work not only reflects upon but also advances strategies toward modernity (primarily in its capitalist modes) and is shaped by the imagination of exceeding its horizon. Such anticipatory modalities of thought frame historical analysis as much as they might comment on the present or project utopian futures. These approaches constitute a distinctive historical consciousness, traces of the past not only being explored from a politicized perspective but also mobilized as purposeful for current situations. For these writers, questions of anticipatory possibility inform their historiography in substantial ways, finding expression explicitly or in the way arguments are pitched—showing even as authors dispute the details of the social narrative at stake, worry about its agents (or their absence), consider the practicality of (or lose confidence in) the future. Even the various "farewells," despite their inevitable tone of reflective disappointment, unavoidably speak from this space.

Although probably one of its most prominent and fertile modes, anticipation is not always utopian. Identifying how utopia functions as an ideology with which to lever capital's future, Manfredo Tafuri advances a cautionary argument, one that emphasizes instead the need to separate the modern movement's correlation with capital from the project of social revolution. Anticipatory politics may take the form of, but is not coextensive with, a narrowly historicist model; those figures considered in *Dialectical Passions* largely avoid it. Fredric Jameson, as we saw, is accused by Hal Foster of holding to a progressivist sense of history, something which can be felt again in Jameson's *A Singular Modernity* where, challenging the notion that there are multiple modernities, he argues that capitalism should be understood as intensifying and expanding as a singularity. Leaving aside the specific arguments at stake here, the very structure of Jameson's historical projection nevertheless involves a forward leaping that partakes of the ruptures associated with the utopian imagination. Even here, "progressivist" does not seem to capture the complexities of Jameson's conceptualizing of temporality, with his concluding demand, which we encountered at the start of this book, that the radical project needs to conduct "archaeologies of the future"—an idea that, I think, is all the more provoking without being limited to the strains of science fiction—precisely because it counters the progressivist conclusions of those melancholic prognoses premised on "forecasts of the past." Whether through Jameson's temporally upturned archaeologies, Chris Marker's interest in making a reversal of Proust with "remembrance of things to come," Foster's deployment of Freudian *Nachträglichkeit*, or Buchloh's efforts to keep a hold on historical memory, radical cultural theory variously contorts its temporal imagination to lay claim to another future.[2] The question of aesthetics and politics is placed back on the agenda, but the political defeatism that undergirds much of this tradition is now less prominent. As should be clear, anticipation is retroactive too, invoking former possibilities and from them projecting other possible outcomes. Benjamin's constellations and dialectical images provide the most well-known models for this collision. Such creative inventions seek to mesh pasts, presents, and futures, not to posit narrowly conceived teleologies, but to make the anticipated focal point the place from which history might recurve to establish different trajectories. Much like avant-garde artists and poets earlier in the twentieth century, possibility in contemporary art and thought is solicited through libidinized energies or by conjuring the transformatory magic of poesis.

Anticipation—while vital and necessary—has nevertheless elicited its own set of difficulties, from "the aestheticization of politics" through to important debates over the problems of "the forcing of the future." I want, however, to pick up on something touched on in the discussion of the Radek Community's actions: the frustration arising from the hiatus of emancipatory politics and the place of art in this difficulty. The intransitivity of possibility and something of its peculiar intersections with art were earlier addressed by Adorno and Lukács, coming to a head in the nineteen-fifties when their intellectual and political differences became starkly posed. Adorno was piqued, to say the least, by his inclusion (and that of Benjamin) alongside those irrationalist philosophers attacked by Lukács in *The Destruction of Reason*; Lukács famously characterized Adorno as having taken up an all-too-comfortable residence in the Grand Hotel Abyss, contemplating the nihil before him. In response, Adorno challenged Lukács's increasing reliance on simple reflection theory, his undialectical faith in immediately given facts, contesting his criticisms of decadent art and distinction between healthy and sick culture.[3] These were all signs, Adorno argued, of Lukács's "stultification" under the Eastern bloc's doctrines of Diamat and socialist realism, and indicated his descent from active critic of to apologist for Stalinist ideology.[4] Adorno's central objection was that Lukács assumed reconciliation to have been achieved in the Soviet Union, but, since social antagonism continued, Adorno called it an "extorted reconciliation" (*Erpresste Versöhnung*).[5] It was exactly this enforced wedding of concept and reality, a false union that concealed social dissonance, which, Adorno noted, Lukács himself had accused absolute idealism of committing.

Given the degree of their mutual antagonism—not to mention the moment's overdetermination by the politics of Stalinism and the opposition of modernism and realism in mid-twentieth-century debates—this conflict of opinions might seem to offer little for considering today's problems. Half a century later, neither of the rival aesthetic models championed has any greater claim on current art, both being grist to the mill of art's ready requisition of historical references (and where, if anything, those deploying social-realist forms—ironically, of course—are often found to be more compelling). The debate between Lukács and Adorno seems anachronistic—and in many ways it is—but if we refuse to rest on its immediate conflictual surface it is also instructive. My aim is not to subject these two figures to some arbitrary posthumous settlement, but to reconsider their differences afresh in order to glimpse some problems for anticipation that still touch on our own moment. What will emerge is in no way an answer (would that answers be so easy to summon)—indeed, the problems will deepen—but, at a point where anticipatory politics are appearing again on the agenda, it seems timely to explore the legacies of an earlier debate and begin to consider how we might to do more than *just* conclude our books and articles with another invocation of future potential. The differences between Adorno and Lukács cannot be brushed aside, but they are often hypostatized, sometimes to the point of falsity. We need, as Adorno himself warns, to avoid relying on those routine little critical *aperçus* to which we are over accustomed. For whatever the stakes of this debate, they are not reducible to being for or against mediation, for or against the Left-Hegelian dialectic of openness, or even simply for or against Hegelian reconciliation—indeed, on these points, we can find a remarkable degree of concurrence between them.

The discussion of possibility emerged around two interwoven issues: how to understand aesthetic realism and the relation to the real in Hegel's thought. In his study of Hegel's early writings, which he positions as the constructive complement to *The Destruction of Reason*, Lukács almost goes so far as to regret Hegel's introduction of the much-debated "reconciliation."[6] Lukács certainly interprets the concept as an index of Hegel's political accommodation and conceives it as "an idealistic mystification of irreconcilable contradictions."[7] Nevertheless, he repeatedly emphasizes that reconciliation is internally contradictory and that it is mistaken to conflate it with Hegel's political conformism. Lukács especially values Hegel's efforts to bring his objective idealism together with his "love of reality."[8] Reconciliation, then, is also interpreted as resulting from the philosopher's "underlying realism, his commitment to the concrete social relations of his age" and his "realization that the contradictions were insoluble."[9] Hegel, he argues, is less interested in the finality of reconciliation (the annulment of contradictions) than in the dialectical "process of annulment."[10] Adorno is more doubtful about Hegel's "excessive confidence in the objective spirit."[11] While he also judges Hegel's dialectics as ultimately serving as "an instrument of identity," he nevertheless insists that it is "a philosophy of identity stretched to the breaking point."[12] "Even the marks of its failure," Adorno maintains, are "struck by truth itself," and rather than just pointing to its weak points, our task is to wrest truth from the all-too-obvious untruths.[13] He challenges those who find in Hegel nothing but a linear progression, dismissing as "one-sided" the perspective "that the movement of the concept is nothing but the advance from one to the other by virtue of the inner mediatedness of the former," criticizing as "ignorant" those who see Hegel's dialectic as "a conceptual straightjacket" or take it to be a "mechanical and coercive" system.[14] Believing its significance to be in his "resistance to the reified consciousness," Adorno draws our attention to the "unresolved and vulnerable quality" of Hegel's philosophy.[15] Thus, of the *Phenomenology*, he notes how, "with incomparable tact, *even the later chapters . . .* refrain from brutally compacting the science of the experience of consciousness and that of human history into one another. The two spheres hover, touching, alongside one another."[16] Adorno also warns against overstating the distinction usually made between the early "radical" and later "conservative" Hegel; a complex mutuality between history and system remains at work and Adorno understands Hegel's "increasing rigidity" and hasty closures as resulting from those specific historical limitations that made experience irreconcilable with the dialectic.[17]

This exchange, however, was not solely concerned with making judgments about Hegel's philosophy; as much as he was the primary object of discussion, "Hegel" also functioned in the debate as a vehicle for addressing more recent intellectual legacies. Lukács challenges what he perceives to be the misuses of reconciliation by the tradition of Neo-Kantianism and Neo-Hegelianism.[18] He objects to the incorporation of Hegel by the "imperialist, reactionary restructuring of neo-Kantianism" in the early twentieth century, which he believes to have exploited the concept in the service of a synthetic national philosophy, blending rational and irrational currents into a single stream, and ultimately into a "Fascistoid neo-Heglianism."[19] Dilthey's 1906 study of Hegel's early writings is taken by Lukács to be foundational for the neo-Hegelian fusion of dialectics with *Lebensphilosophie*, and on which, he complains, even Marxist scholarship had become dependent.[20] Adorno's tack is different. We should not sift Hegel's writings to separate what might be deemed to be alive from what was dead (an approach he attributes to Benedetto Croce); we should ask, rather, "what the present means in the face of Hegel."[21] By this he means us to reflect on the disjunction between a promise set forth—what Adorno means when he asks us to confront Hegel's truth content—and the reality before us (an imperative similar to the insistence in recent radical political philosophy that we face up to Marx's "actuality").[22] The burden of relevance, then, presses not on the past but on the present. Rather than repeating such truisms as "Hegel belongs to the early nineteenth century" or that "today is a different world," demand instead that the present answer for its failure to live up to a potential once anticipated. With this approach, Adorno challenges the complacent assumption of intellectuals who believe themselves to have "moved on" (one which was then based on the belief in reason's progress, but which, we might note, remains intact through those later intellectual discourses that doubt progression). Are we not becoming, he wonders, "accommodated to what merely exists"?[23] This collapse into identity (our reconciliation with capital) also harbors a moment of nonidentity: contemporary thinkers can believe they "are above something," Adorno argues, but "only because they are not in it."[24]

Both Adorno and Lukács, then, saw reconciliation in Hegel as an outcome of a confrontation with material irreconcilability, although there was a slight difference in how they emphasized this. According to Adorno, Hegel is unable to reconcile his dialectic with his experience of social reality, identifying "no real historical force on the other side of" bourgeois society; for Lukács, the problem of irreconcilability arises from Hegel's commitment to "social reality as it actually is."[25] Their positions are very close, but the gap between their descriptions is indicative of the central issue to emerge from their debate and which is rather more subtle than their opposition in many analyses can allow. The problem they grapple with is how to think through the material and processual. As Adorno insists against Dilthey, dialectical processes are not just dynamic, but "a dynamic of fixed and dynamic elements."[26] The vitalist appeal to process turns an ideological blind eye to the "unreconciled epoch," offering an essentially smooth conception that—unlike dialectics' acknowledgement of discontinuity and willingness to think "in and through the extremes"—evades confronting capitalism's "antagonistic totality."[27] The question being addressed by Adorno and Lukács, then, was how to grasp the relation between material reality and its processual consideration; and, as a variant on this question, how to make the transition from social reality "as is" to "as it could be." Reflection on this relation between material reality, material processes, and a material-processual understanding underpins the familial connection between anticipation and reconciliation.[28] With the appeal to possibility, one of the central planks for today's Left intelligentsia, this debate over the distinction between real and abstract possibility seems crucial to reconsider. Radical thought—whether in art, or in aesthetic or political theory—is still haunted by the gulf between the present reality and an unrealized promise. From this perspective, "openness" does not figure freedom but serves instead as a harsh reminder of its absence. The struggle with this contradiction (or, for some, paradox) provides the contours of this exchange.

It is significant that, in the Adorno-Lukács debate, an essentially political dispute is conducted almost entirely on the terrain of literature—a displacement that is neither just an allegory for, nor simply a sublimation of, politics but which has some peculiar repercussions for the argument itself. It is worth looking more closely at this feature of their debate. Lukács criticizes modern literature for its subjectivism and what he sees as its abstract attachment to possibility.[29] He believes realism, in contrast, to be able to posit possibility concretely because it employs, in his view, a dialectic of subjective and objective factors. Mistaking potential for reality, modernist writing succumbs to melancholy when it finds that the world refuses to accord with those imagined possibilities. Contrary to common assumptions, while Adorno disdains socialist realism and even subjects Lukács's "critical realism" to devastating critique, he does not reject realism as such. Instead, he advances an account which, firstly, considers realism at the levels of both content and form (and aiming at their fusion), and, secondly, understands realism not as a "reflection" but as a transfiguration of the given. The strength of possibility in the work of art, he argues, "depends on the degree of reality it has absorbed into itself, no matter how transformed and reconfigured that reality may be."[30] Adorno reconsiders the standard contrast of realist and modernist literature in a series of quite brilliant observations in which some of Lukács's favored examples are subjected to reversals: Balzac, for instance, becomes less realistic, Proust more so. Balzacian realism is "derived" rather than "primary," Adorno argues. Far from "yielding to realities," Balzac deforms their façades and "stares them in the face until they become transparent down to their horrors."[31] Concreteness here is "excessive"; the exaggerated accumulation of descriptions of places and characters are comparable to "restitution phenomena," in which details of the "lost objects" of reified reality are sketched. In a letter to Thomas

Mann, which considers the relation of subjectivity to the demands of realism and of external obligation to aesthetic reason, Adorno describes his insight while reading *La recherche* (and which has echoes not only of the moment when Marcel recognizes the true force of Giotto's *Charity* but also of Benjamin's entry in his Moscow diary). To his encounter with Proust, Adorno attributes the realization that "the greater the precision one maintains with regard to the historical details . . . the more likely one is to achieve spiritualization and attain the world of the *imago*."[32] Elsewhere, challenging Lukács directly, and again discussing Proust, he notes how "the most precise 'realistic' observation is . . . intimately connected with the formal aesthetic law."[33] Lukács, Adorno argues, abuses Hegel's distinction between concrete and abstract possibility in order to uphold a traditionalist aesthetic. More precisely, Adorno objects to Lukács's use of these two senses of possibility to sustain an unreflexive form of realism that pits "pragmatic fidelity" against unrealistic method.[34]

But there are questions to be asked of the way these two types of possibility are discussed. Lukács's insights into different tendencies in literature descend into the dichotomy of realism and modernism, in which modernism fails to distinguish abstract and real potential, and where the abstract potentiality of modernist literature leads to an attenuation of reality and dissolution of personality. These results, along with the melancholy Lukács attributes to modernism's experience of the disjunction between reality and imagined potential, imply real political problems. Of course, contrary to Adorno's accusations, Lukács himself does not adhere to a reflectionist model of realism, passive mimesis being closer to the naturalism with which realism is contrasted. For Lukács too realism is to be understood as transfigurative, but he conceives this transformative role differently from Adorno, and he sees the ethical exempla literature provides as the primary means for achieving such transition. However, by conducting his argument in relation to literature, some specific outcomes surface. Possibility is located in terms of plot development and in the perceived coherence of the motivations, decisions, and choices of literary characters. Yet, development, coherence, and motivation here are inseparable from the qualities of a particular literary mode. Artistic form determines whether concrete possibilities become concretized or whether possibilities remain intransitive and thus abstract; an authored denouement retrospectively legitimizes (or delegitimizes) the outcome for the various potential developments introduced. There is a disjunction, then, between the inner structural coherence of the novels praised by Lukács (upon which his argument turns), and the political stakes he invokes, which concern the way abstract and real possibility might be worked through in an extra-aesthetic realm. (One of the interesting things about much radical art practices of the recent period is that even when clearly situated as practices of critical autonomy, it is precisely this extra-aesthetic effectivity which provides—and I mean this not casually but substantially—the strong animating force for their work.)

Adorno is much more explicit (even immoderate) in claiming powers for aesthetic experience, yet at the same time his argument—whether it be pitched to reality or to utopia—makes more direct political appeals. It is as though he sets out to deploy, rhetorically, an excess of both aesthetics and politics. Although much of his discussion addresses literature, his sense of temporality is fundamentally shaped by musical structures. This is articulated well in his explorations of Hegel's philosophy, which he compares to the "retroactive" force of Beethoven's recapitulation and where the "advance is a permanent critique of what has come before."[35] His point is also made with less elevated examples and, interestingly, is treated as an embodied praxis. Hegel's prose is characterized by its physical awkwardness, or the "gestural or curvilinear" qualities of his writing. The reader is "borne by the current," yet must also acquire "an intellectual slow-motion procedure."[36] For Adorno, this combination of temporality and plasticity—with its corporeal qualities, its "antilinguistic impulse," the way it "eludes literary presentation," or is "more like films of thought than texts"—offers a basis for "a different kind of understanding."[37] Furthermore, he emphasizes a point that Lukács acknowledges briefly but fails to develop: that transition can be made from abstract to real possibility. Adorno also defends abstract possibility by appealing to the transfigurative role of utopian reach. Although the movements described are essentially aesthetic, he does not make his case through the example of literature or music but makes it politically:

> Even the idea that opposes reality in holding fast to a possibility that is repeatedly defeated does so only by regarding that possibility from the point of view of its realization, as a possibility for reality, something that reality itself, however weakly, is putting

> out feelers to, and not something that "would be nice," the tone of which resigns itself to failure from the outset.[38]

Significantly, reaching out is not posed as a provisional desire (which here anticipates disappointment, echoing Lukács's concern over melancholy). Rather, it is an extension predicated in extant tendencies. "Possibility," then, is conceived retroactively from the perspective of its realization, but simultaneously antennae in the present turn to sense and touch this future.

What happened to Adorno's argument in subsequent years helps illuminate the distinctive intersection of plasticity and temporality described thus far. In *Aesthetic Theory*, he argues that art "wills the other," yet in his writings from the fifties, elements of this other are already present in reality, needing to be displaced into new constellations to liberate transfigurative potential. But in *Aesthetic Theory* Adorno's account seems to slacken its former hold on the reflexive and transfigurative conception of realism (indeed, realism seems reduced and, rather than being contested, is increasingly consigned to "copy theory"). What continues—and becomes more exaggerated—is the emphatic and direct articulation of both art and politics. The immediacy with which political expression is uttered has certain advantages over Lukács's account but it also harbors significant drawbacks for Adorno, and these come to the surface once his attention to the "possibility for reality" subsides. Art testifies to an "other" state of things just by positing "the actuality of the non-existent"; art "is the possible, as promised by its impossibility."[39] This is not a new idea for Adorno to express, but the context of this idea changes. Also noticeable—resulting, in part, from the exigencies of expositing an aesthetic theory—is the relative loss of detail still animating his earlier discussions. "Art" as such displaces what he formerly achieves by deploying specific artistic devices, strategies, and modalities. His conception of aesthetics and his conception of politics become more abstracted, his presentation of them more declarative, while their relation to one another is increasingly given by direct assertion. This problem shows up dramatically in Adorno's approach to describing art's utopic reach. Art deploys remembrance, he argues, "to give flesh and blood to the notion of Utopia, without betraying it to empirical life."[40] The materiality of the present no longer contains possibility, however weak; posed now as mere "empirical life," the present is conceived entirely in its reified aspects. Adorno's metaphors of corporeality disappear; the feelers, the reaching out, the attempts to touch or to find footholds, are evacuated from current reality. Utopia too remains fleshless and bloodless unless it can tap the body of remembrance. The latter is entrusted to be bearer of nonreified life, but its relation to the future becomes increasingly disjunctive; it is only able to connect by fiat, and, at the same time, the very appeal to recollection struggles to span this temporal fissure. At this point, we might, I suppose, consider appealing to the power of constellatory thought, but, unlike its Benjaminian model, Adorno's version seems increasingly powerless to spark political-historical poesis. Adorno's methodological aspiration to "pass through the extremes" begins to give way to its opposite: a failure to pass through, becoming more akin to an exaggeration of, and a paralysis in, these extremities.

Concerns over the distinction between abstract and real possibility have a longer history in Lukács's thought, playing a particularly important role in the years preceding and following the Hungarian republic. Through this period Lukács reappraised the tragic outlook that he had initially adopted from Neo-Kantianism and that he subsequently refashioned into revolutionary romanticism—"the unbridgeable chasm between the reality that is and the ideal that should be."[41] He was also concerned with the limitations of utopian desire and of the attempts to leap from "is" to "ought" (*Sollen*) through voluntarist actions—a question that became sharpened with the defeat of the March Action in Germany of 1921, (an event in which the encouragement of Karl Radek, namesake of the artists' collective, had played a central role). In addition to advancing his account of reification, Lukács's *History and Class Consciousness* can be taken as a theoretical reckoning with the events of 1921.[42] Hegel's discussion of objective possibility played an important role for Lukács's deliberations. Yet, despite this significant reorientation, according to István Mézsáros, this "unresolved duality" of "is" and "ought" continued to shape Lukács's ontology.[43] Rejecting the obligation proposed by individual ethics, his approach remained based on an "imperatival foundation," orchestrated around autonomous "decisions between alternatives."[44] From *History and Class Consciousness* on, Lukács substituted "ought" with "concretize." For Mézsáros, however, while the *Zweispalt* was "given a dialectical, and increasingly more concrete, assessment," the concrete remained a category at the level of a general philosophical synthesis and lacked practical orientation.[45] In Mézsáros's view, Lukács later found a way to advance his exploration of totality and mediation in the context of aesthetics. Yet, as I have noted, the writings on art can also be seen as playing out these same contradictions: naturalism/realism; realism/modernism; narration/description; reportage/portrayal. And in these discussions, with Lukács's aesthetics serving as a political medium, appear the familiar dualities of society versus the individual, structure versus agency, objective or subjective, spontaneous causality versus intentional causality, decisions to act or not act.

Adorno famously replaced Hegel's "identity of identity and nonidentity" with the "nonidentity of identity and nonidentity," hoping thereby to avoid prematurely reconciling the dialectic. His deliberate reformulation, however, is too often represented as the supplanting of identity by nonidentity, and so, in those numerous confrontations staged to unseat the dialectic, we end up with the installation of a counterdoxa of abstract negations and fragments. It is not just dialectics that stalls, however; constellatory thought registers the same hiatus. In the end both dialectics and constellations falter for reasons less elevated than their internal intellectual frames, and for reasons that are more concrete and elusive. The problem is also manifested in the political arena with the appeals to various jumps—Benjamin's "tiger's leap," the interest in the "Pascalian wager," or even in "will," or Schmittian decisionism.

The problem of connecting possibility's modes—the link between potential and real, to make their relation transitive, the paralysis in contraries—has simmered through *Dialectical Passions*: the rigid petrification of allegorical forms and allegory's postmodern simplifications; dialectics at a standstill; the hobbyhorse in Jackson Pollock's painting and T. J. Clark's decision to have the wooden head stand as antinomy because the means to reconcile capital's reified dualities dialectically do not exist.[46] Another variant on the problem can be found in those positions that eradicate tension altogether, suppressing one side of their duality to project a social one-dimensionality and the end of critical distance. In the context of these debates, concerns over the nature or deployment of tropes are never simply internal questions of close reading. Metaphor, homology, analogy, affinity, and association become these authors' means to break through reification and to grasp the manifestation of social mediation in thought or in poetic or methodological practice. Essentially antinihilist in intent, Clark's drive to establish mediation ran into aporia, the failure to grasp the social reality of the sign. Tafuri's historiography specifically tried to embrace rather than reject modernity's nihilism, hoping thereby to break the debilitations of tragic politics.[47] Yet, despite the emphatic distrust of dialectical mediation in the thought of Tafuri and others, their venture may, paradoxically, have come closer to touching it. Situated in a context where transitional social forms were being contested, for a period at least this project could develop the sense of a nonabstracted possibility—such that "utopia" could be dismissed as a barrier to emancipatory praxis. Shorn of "feelers," toying with the same Left-nihilist philosophy—even when predicated on radical intentions—acquires a rather different cast. Jameson modified Tafuri's position, recombining utopian models into the antitragic argument (the refusal of backward-looking sentimentalism characterizing Jameson's approach). Again, what was lost was a context from which this antitragic position, however debatable and controversial it had been in its own terms, had once made sense. Interestingly, where Jameson sees in the architecture of late capitalism hyper-urban extensions, Tafuri detects antiurban retreats; the abstractions imagined here are different. The loss of a center for Jameson increasingly became the "loss of reality" (first identified as a societal tendency and then ontologized).

Classic appraisals of Western Marxism from the New Left, such as Perry Anderson's, drew attention to the ways intellectual disillusion with praxis and the growth of social pessimism among radicals had underpinned the turn to aesthetics. Preceding the onset of the long nineteen-eighties, this moment was that of the New Left's own crisis. Today's revived conjunction of aesthetics and politics is not a mere continuation of these preceding formations. Not that optimism, or even social pragmatism, has suddenly taken hold, but the mood of the intervening years has fissured and one hears calls for answers made with increasing urgency. Indeed, what we see are less retreats from than advances toward politics. It would be wrong to conclude, however, that the politicization in art means that art is the new site of political action; there may well be political effects, but these efforts mostly remain circumscribed on art's terrain. We need to be more cautious in how we understand this politicization. The intensity of this sublimation unleashes immense creativity and productivity, but shorn of a renewal in unified social practice the repoliticization of aesthetics in the twenty-first century has probably achieved all it can, though it often achieves this by pushing at the very limits of its own enclosures. This is not to say that there is nothing more to be thought or debated about art, or about its relation to society and to projects of social emancipation; but it is symptomatic of the current juncture that among theorists and practitioners of art we hear posed once again the question—now over a hundred and fifty years old—"What is to be done?"[48]

Notes

Introduction

1. Chris Marker, dir., *Sans soleil* [*Sunless*], DVD, directed by Chris Marker (Paris: Argos Films, 1982).

2. William Morris, "A Dream of John Ball," in *Three Works by William Morris* (London: Lawrence and Wishart, 1968), 53.

3. E. P. Thompson, *The Poverty of Theory and Other Essays* (London: Merlin, 1978), 280; Perry Anderson, *Arguments within English Marxism* (London: Verso/NLB, 1980), 17 (the context is Anderson's consideration of Thompson's critique of Louis Althusser); epigraph to Michael Hardt and Antonio Negri, *Empire* (Cambridge, MA: Harvard University Press, 2000).

4. Some important points of reference include Fredric Jameson, *Marxism and Form: Twentieth-century Dialectical Theories of Literature* (Princeton: Princeton University Press, 1971); Martin Jay, *Dialectical Imagination: A History of the Frankfurt School and the Institute of Social Research 1923–1950* (London: Heinemann Education Books, 1973); Ernst Bloch et al., *Aesthetics and Politics* (London: New Left Books, 1977); Susan Buck-Morss, *The Origin of the Negative Dialectics: Theodor W. Adorno, Walter Benjamin and the Frankfurt Institute* (New York: Free Press, 1977); Gillian Rose, *The Melancholy Science: An Introduction to the Thought of Theodor W. Adorno* (London: Macmillan Press, 1978); Eugene Lunn, *Marxism and Modernism: An Historical Study of Lukács, Brecht, Benjamin, and Adorno* (Berkeley: University of California Press, 1982); Martin Jay, *Marxism and Totality: The Adventures of a Concept from Lukács to Habermas* (Berkeley: University of California Press, 1984).

5. Roy Bhaskar, *Dialectic: The Pulse of Freedom* (London: Verso, 1993).

6. Theodor Adorno, "Skoteinos, or How to Read Hegel" (1962–1963), in *Hegel: Three Studies*, trans. Shierry Weber Nicholson (1963; Cambridge, MA: MIT Press, 1993), 146

7. Adorno, "The Experiential Content of Hegel's Philosophy" (1959), in *Hegel*, 56.

8. Johann Wolfgang von Goethe, "Prologue from Heaven," in *Faust*, lines 294–295.

9. Theodor Adorno, *Aesthetic Theory*, trans. Robert Hullot-Kentor (London: Athlone, 1999), 66. Although Adorno's discussion concerns natural beauty, I sometimes wonder whether what he had in mind was not so much avine sounds as their simulation by modernist composers such as the devout Catholic and organist at La Trinité, Olivier Messiaen. From the fifties until the end of his life, Messiaen transposed bird sounds into musical notation.

10. See, for example, Robert Storr's opposition to Yve-Alain Bois over the question of Robert Ryman: Robert Storr, "Simple Gifts," in *Robert Ryman* (London: Tate Gallery Publications, 1993); Yve-Alain Bois, "Ryman's Tact" (1981), in *Painting as Model* (Cambridge, MA: MIT, 1990), 215–226.

11. Gustav Metzger, *Auto-Destructive Art* (London: Destruction/Creation, 1965); Eva Hesse in Germano Celant, *Arte Povera: Conceptual, Actual or Impossible Art?* (London: Studio Vista, 1969), 56; essays by Robert Smithson republished in *The Writings of Robert Smithson*, ed. Nancy Holt (New York: New York University Press, 1979).

12. Alighiero Boetti in Germano Celant, *Arte Povera,* 156.

13. Carl Andre in Germano Celant, *Arte Povera*, 204.

14. Barbara Rose, "ABC Art," *Art in America* (October-November, 1965); Donald Judd, "Specific Objects," *Arts Yearbook* 8 (1965), and republished in *Donald Judd: Complete Writings 1959–1975* (Halifax: Nova Scotia College of Art and Design/New York University Press, 1975), 181–189.

15. Terry Atkinson, "Disaffirmation and Negation," *Mute 1* (Copenhagen: Galleri Prag, 1988), 8–13.

16. See, for example, *The Philistine Controversy*, ed. Dave Beech and John Roberts (London: Verso, 2002).

17. T. J. Clark, "Jackson Pollock's Abstraction," in *Reconstructing Modernism: Art in New York, Paris, and Montreal, 1945–1964*, ed. Serge Guilbaut (Cambridge, MA: MIT, 1990), 172–173; the theme of negation is flattened out for the revised version in *Farewell to an Idea: Episodes from a History of Modernism* (New Haven: Yale University Press, 1999), 299–300.

18. Michael Fried, "Three American Painters: Noland, Olitski, Stella" (1965), in *Art and Objecthood: Essays and Reviews* (Chicago: University of Chicago Press, 1998), 214.

19. Benjamin quotes Brecht, who in turn had rephrased a point that he attributes to Sternberg. Bertolt Brecht, "No Insight through Photography"

(c. 1930) and "The *Threepenny* Lawsuit" (1931), *Bertolt Brecht on Film and Radio*, ed. and trans. Marc Silberman (London: Methuen, 2000), 144 and 164. Walter Benjamin, "Little History of Photography" (1931), *Walter Benjamin: Selected Writings*, Volume 2: 1927–1934, trans. Edmund Jephcott and Kingsley Shorter (Cambridge, MA: The Belknap Press of Harvard University Press, 1999), 526.

20. Charles Harrison, *Essays on Art & Language* (Oxford: Basil Blackwell, 1991), 232.

21. Laura Mulvey, "Changes: Thoughts on Myth, Narrative and Historical Experience," *History Workshop Journal* 23 (1987): 3–19. What Mulvey has in mind are practices of negation of a radical or critical kind: cultural politics, art practice with serious political intentions (and there are, of course, more strictly political variations of this argument).

22. See, for example, Sheila Fitzpatrick, *The Russian Revolution* (Oxford: Oxford University Press, 1994), especially the section "Interpreting the Revolution" in the Introduction to the Second Edition. Fitzpatrick does not specifically address the term "negation," but the parallels in argumentative form are striking.

23. Harrison, *Essays*, 232.

24. Max Horkheimer, foreword to Martin Jay, *Dialectical Imagination: A History of the Frankfurt School and the Institute of Social Research 1923–1950* (London: Heinemann Education Books, 1973), xi. See also Max Horkheimer, "Traditional and Critical Theory" (1937), *Critical Theory: Selected Essays*, trans. Matthew J. O'Connell et al. (New York: Herder & Herder, 1972).

25. Herbert Marcuse, *Negations: Essays in Critical Theory*, trans. Jeremy J. Shapiro (1968; repr. London: Free Association Books, 1988), xi.

26. Giorgio Agamben, *Language and Death: The Place of Negativity*, trans. Karen E. Pinkus with Michael Hardt (1982; repr. Minneapolis: University of Minnesota Press, 1991). Massimo Cacciari, *The Necessary Angel*, trans. Miguel E. Vatter (1984; repr. Albany: State of New York University Press, 1994).

27. T. J. Clark, note added to the republication of "Clement Greenberg's Theory of Art" in *Pollock and After: The Critical Debate*, ed. Francis Frascina (London: Harper and Row, 1985), 55.

28. Rosenberg had in mind the "revolutionary event in Aesthetics" in Susanne Langer's *Feeling and Form* (also, no doubt, he intended a nod toward Marx's "Eleventh Thesis." Harold Rosenberg, "Virtual Revolution," in *The Tradition of the New* (1959; repr. Chicago: University of Chicago Press, 1982), 56.

29. Since the publication of Hauser's key works, the term "sociological" has often acquired a derogatory ring in discussions of art. His understanding of the "sociological approach" is very different to the sociological models proposed by more recent writers such as Pierre Bourdieu or Janet Woolf.

30. Max Raphael, "Zur Kunsttheorie des dialektschen Materialismus" (1932), in *Proudhon, Marx, Picasso: Three Essays in Marxist Aesthetics* (1933; repr. London: Lawrence & Wishart, 1980); a section of *Zur Erkenntnistheorie der konkreten Dialektik* (1934), appeared as "A Marxist Critique of Thomism," *Marxist Quarterly* 1, no. 2 (1937): 285–292. For an introduction to the art-historical figures associated Marxism, see Andrew Hemingway, ed., *Marxism and the History of Art: From William Morris to the New Left* (London and Ann Arbor: Pluto, 2006).

31. Frederick Antal, "Remarks on the method of art-history" (1949), in *Classicism and Romanticism, with Other Studies in Art History* (London: Harper and Row, 1973), 175–189. In additions to specifically art-historical texts (Warburg scholarship, Schapiro's essays on Silos and the Ruthwell Cross, Blunt's book on Italian art theory from 1940, and Millard Meiss), Antal mentions G. M. Trevelyan's *English Social History* (1942) and George Thomson's *Aeschylus and Athens* (1941). Writing to Lukács just a couple of years earlier, Antal claimed that Lukács's writings remained "closer to me than all the American or English books in the last few years." Antal cited Arpad Kadarkay, *Georg Lukács: Life, Thought, and Politics* (Cambridge, MA: Blackwell, 1991), 368.

32. Perry Anderson, "Components of the National Culture," *New Left Review* 50 (1968); revised version in Anderson, *English Questions* (London: Verso, 1992), 48–104.

33. For a critical assessment of Warburg studies since the New Left, see Otto Karl Werckmeister, "The Turn from Marx to Warburg in West German Art History, 1968–90," in *Marxism and the History of Art*, ed. Andrew Hemingway, 213–220.

34. Two more art historians, Károly (Charles de) Tolnay and Johannes Wilde, were also involved in the Circle.

35. Georg Lukacs, *Record of a Life: An Autobiographical Sketch* (London: Verso, 1983), 51; Balázs cited Lee Congdon, *Exile and Social Thought: Hungarian Intellectuals in Germany and Austria 1919–1933* (Princeton: Princeton University Press, 1991), 52.

36. *Arnold Hauser Im Gespräch mit Georg Lukács* (Munich: Verlag C. H. Beck, 1978); an English translation of the radio interview was published as "On Youth, Art and Philosophy," *The New Hungarian Quarterly* 16, no. 58 (1975).

37. Walter Benjamin, "Rigorous Study of Art: On the First Volume of *Kunstwissenschaftliche Forschungen*" (1931/33), in *The Vienna School Reader: Politics and Art Historical Method in the 1930s*, ed. Christopher S. Wood (New York: Zone Books, 2000), 439–451.

38. Meyer Schapiro, "The New Viennese School" (1936) in *The Vienna School Reader: Politics and Art Historical Method in the 1930s*, ed. Christopher S. Wood, 452–485; and for the critique of Barr see "The Nature of Abstract Art."

39. See, for example, Antal, "Remarks on the Method of Art History."

40. Antal saw the Vienna School (Riegl, Wickhoff, Dvořák) as more historically inclined than Wolfflinian art history and reported how Dvořák insisted that Riegl's method was characterized not by "absolute aesthetics" but by "psychological and historical conceptions of art-history" (Antal, "Remarks on the Method of Art History," 176). Antal cited Warburg as the other strand of new methods in art history, praising Warburg's rescue of art from aesthetic isolation, his emphasis on art's cultural-historical context, and his refusal to assume barriers between high art and popular forms. Warburg's own project can be seen as either *Kunstwissenschaft* or *Kulturwissenschaft*. See Michael Ann Holly, *Panofsky and the Foundations of Art History* (Ithaca: Cornell University Press, 1984), 104ff.

41. Frederick Antal, *Florentine Painting and Its Social Background: The Bourgeois Republic Before Cosimo de' Medici's Advent to Power: 14th and Early 15th Centuries* (London: Routledge and Kegan Paul, 1948).

42. Hauser, *Philosophy of Art History* (London: Routledge and Kegan Paul, 1959), 137. Aloïs Riegl and his notion of *Kunstwollen* represented for Hauser the opposite pole within a school of thought that derived its approaches from nineteenth-century philosophy of history. Assessing historicism in its wider form, Hauser noted how: "It adopts the mystifying method of referring every historical event to some superindividual—ideal, divine, or primeval—origin, but combines with this an individualizing treatment that asserts not simply the uniqueness, but also the absolute incomparability of historical structures, and so concludes that every historical achievement, and thus every art-style, must be measured only against its own acknowledged standards." (Hauser, *Philosophy of Art History*, 120).

43. Nicos Hadjinicolaou, *Art History and Class Struggle*, trans. Louise Asmal (1973; repr. London: Pluto Press, 1978). *Kritische Berichte*, 4, nos. 2 and 3 (1976).

44. On these and other developments, see Andrew Hemingway, "New Left Art History's International," in *Marxism and the History of Art*, 175–195.

45. See, for example, *The New Art History*, ed. A. L. Rees and F. Borzello (London: Camden Press, 1986); Lisa Tickner and Jon Bird, eds., introduction to *The Block Reader in Visual Culture* (London: Routledge, 1996).

46. The sociological work of Pierre Bourdieu has been influential in this regard.

47. At the conference of the German Sociological Association in Tübingen in 1961, the argument between Karl Popper and Adorno on the logic of the social sciences initiated the dispute between critical rationalism and critical theory. Although widely used to describe the Popperian position, "logical positivism" can be too loosely applied: Popper himself opposed the logical positivists of the Vienna Circle, and most of the "positivists" in the dispute disclaimed the designation. See David Frisby, "Introduction to the English Translation"

of Theodor Adorno et al., *The Positivist Dispute in German Sociology*, trans. Glyn Adey and David Frisby (1969; repr. London: Heinemann, 1976), ix, x–xi. Frisby (xv) draws his discussion of theoretical incommensurability from Paul Feyerabend's "Against Method," *Minnesota Studies in the Philosophy of Science* 4 (1970), eds. M. Radnor and S. Winokur; the argument was expanded in *Against Method: Outline of an Anarchist Theory of Knowledge* (London: New Left Books, 1975). Horkheimer's arguments were advanced in "Traditional and Critical Theory" (see also "The Latest Attack on Metaphysics").

48. Ernst Gombrich, "The Social History of Art" (1953), in *Meditations on a Hobby Horse and Other Essays on the Theory of Art* (London: Phaidon, 1963), 86, 88.

49. Ibid., 88.

50. Ibid.

51. Ibid., 89–90.

52. Ibid., 91, 92. While distancing himself from Hauser's orthodox Marxism and regretting his resort to "cant terms of Marxism" (such as "dialectic"), Greenberg was far more generous in his 1951 review than Gombrich would be; indeed, his judgment is the very opposite. Hauser, Greenberg writes, "respects the manifoldness of his vast subject, rides no theses and tries not to simplify in order to synopsize" (95); "he remains both an art critic and sociologist." Recognizing that Hauser's approach was not reducible to vulgarized versions of Marxism even if it did sometimes sink to "glib equations" (95), Greenberg praises Hauser's sensitivity to aesthetic questions (in part because Hauser did not set out to develop a coherent Marxist aesthetics); his attempts to develop social explanations not only for artistic content but also for form; his ability to "do justice to the difficult complexity" of relations of social and artistic phenomena (especially for nineteenth-century art); and his refusal to sacrifice art's complexity at the alter of mass accessibility (97–98). Clement Greenberg, "Review of *The Social History of Art* by Arnold Hauser" (1951), *The Collected Essays and Criticism*, Volume 4, ed. John O'Brian (Chicago: University of Chicago Press, 1993), 94–98.

53. Gombrich, "Social History of Art," 93.

54. I discuss this in my review of Matthew Beaumont et al., eds., *As Radical as Reality Itself: Essays on Marxism and Art for the 21st Century* and Andrew Hemingway, ed., *Marxism and the History of Art: From William Morris to the New Left* in Gail Day, "A real surprise," *Radical Philosophy* 149 (2009): 59–62.

55. Gilles Deleuze, *Nietzsche and Philosophy*, trans. Hugh Tomlinson (1962; repr. London: Athlone, 1983).

56. One of the more sustained versions of this argument can be found in Andrew Benjamin's work. His opposition to the *via negativa* in his consideration of modern art responds directly to the force of Adornian tendencies. Andrew Benjamin, *Object Painting* (London: Academy Editions, 1994).

57. Karl Marx, "Afterword to the Second German Edition" (1873), *Capital*, vol. 1, 29. See also the essays in *Lenin Reloaded: Towards a Politics of Truth*, ed. Sebastian Budgen, Stathis Kouvelakis, and Slavoj Žižek (Durham: Duke University Press, 2007) and especially Kouvelakis's contribution, "Lenin as Reader of Hegel: Hypotheses for a Reading of Lenin's Notebooks on Hegel's *The Science of Logic*," 164–204.

58. Karl Marx, "Economic and Philosophical Manuscripts" (1844), *Early Writings*, trans. Rodney Livingstone and Gregor Benton (Harmondsworth: Penguin Books, 1975), 386.

59. Marx, *Capital*, vol. 1, 29.

60. David McLellan, *The Young Hegelians and Karl Marx* (1969; repr. London and Basingstoke, 1980), 19; Jacques Derrida, "From Restricted to General Economy: A Hegelianism without Reserve," *Writing and Difference*, trans. Alan Bass (1967; repr. London: Routledge and Kegan Paul, 1978).

61. Arnold Ruge, *Zwei Jahre in Paris* (1846), cited McLellan, 20; Marx, *Capital*, vol. 1, 29.

62. August von Cieszkowski, "Teleology of World History," in *Prolegomena to Historiosophie* (1838), trans. André Liebich, in *The Young Hegelians: An Anthology*, ed. Lawrence S. Stepelevich (Cambridge: Cambridge University Press, 1983), 57–89.

63. McLellan, 18.

64. Edgar Bauer, "The Political Revolution," *Critique's Quarrel with Church and State* (1844), trans. Eric von der Luft, in Stepelevich, *Young Hegelians*, 265, 266, 267.

65. G. W. F. Hegel, *Phenomenology of Spirit*, trans. A.V. Miller (Oxford: Oxford University Press, 1977), 10 (the citations that follow are from 9–11).

66. Ibid., 29ff., 35, 41–2.

67. Ibid., 43.

68. Adorno, *Hegel*, 62.

69. Hegel, *Phenomenology*, 17–18.

70. Ibid., 23.

71. Ibid., 21.

72. Ibid., 18–19.

73. Jean-Luc Nancy, *Hegel: The Restlessness of the Negative* (Minneapolis: University of Minnesota Press, 2002), 3.

74. Perry Anderson, *Considerations on Western Marxism* (London: New Left Books, 1976).

75. Ernst Bloch et al., *Aesthetics and Politics* (London: New Left Books, 1977).

76. J. M Bernstein, *The Fate of Art: Aesthetic Alienation from Kant to Derrida and Adorno* (Cambridge: Polity Press, 1992); Andrew Bowie, *Aesthetics and Subjectivity: From Kant to Nietzsche* (Manchester: Manchester University

Press, 1990); and Bowie, *From Romanticism to Critical Theory: The Philosophy of German Literary Theory* (London: Routledge, 1997); Terry Eagleton, *The Ideology of the Aesthetic* (Oxford: Blackwell, 1990).

77. Gillian Rose, *Dialectic of Nihilism: Poststructuralism and Law* (Oxford: Blackwell, 1984); Peter Dews, *The Logics of Disintegration: Poststructuralist Thought and the Claims of Critical Theory* (London: Verso, 1987); Robert Hullot-Kentor's essays collected in *Things Beyond Resemblance: Collected Essays on Theodor W. Adorno* (New York: Columbia University Press, 2006). See also Peter Osborne's essays in the journal *Radical Philosophy*, his contributions (one coauthored with Peter Dews) to the collection *The Problems of Modernity: Adorno and Benjamin*, ed. Andrew Benjamin (London: Routledge, 1989), and Osborne, *Philosophy in Cultural Theory* (London: Routledge, 2000). *The Philistine Controversy* provides an interesting footnote to these developments with specific reference to debates in contemporary art, drawing together debates from *New Left Review* in the late nineties with new contributions, and also includes my own intervention, "When Art Works Crack(le)." Recent work drawing on Walter Benjamin includes: Daniel Bensaïd, *Marx for Our Times: Adventures and Misadventures of a Critique*, trans. Gregory Elliott (1995; repr. London: Verso, 2002); Michael Löwy, *Fire Alarm: Reading Walter Benjamin's On the Concept of History*, trans. Chris Turner (2001; repr. London: Verso, 2005).

78. Giorgio Agamben, *The Man without Content*, trans. Georgia Albert (1970; repr. Stanford: Stanford University Press, 1999).

79. Jacques Rancière, "The Aesthetic Revolution and its Outcomes." *New Left Review* 14 (2002): 133–51.

80. Alain Badiou, *Handbook of Inaesthetics*, trans. Alberto Toscano (1998; repr. Stanford: Stanford University Press, 2005), 5.

81. Jacques Rancière, *The Politics of Aesthetics: The Distribution of the Sensible*, trans. Gabriel Rockhill (2000; repr. London: Continuum, 2004), 24.

82. Fredric Jameson, *A Singular Modernity: Essay on the Ontology of the Present* (London: Verso, 2002), 215. The point had been made before, but, by and large, not heard. István Mészáros has argued that the discussions of modernity have masked the critical analysis of capitalism. István Mészáros, *The Power of Ideology* (New York: Harvester Wheatsheaf, 1989).

83. Jameson, *A Singular Modernity*, 215.

84. Gopal Balakrishnan, ed., *Debating Empire* (London: Verso, 2003).

85. Bensaïd, *Marx for Our Times*, 2.

86. Editorial statement to first issue of *Historical Materialism: Research in Critical Marxist Theory*, 1997:1.

87. Alex Callinicos, *An Anti-Capitalist Manifesto* (Polity Press: Cambridge, 2003). Critical commentary by Naomi Klein, especially her book *No Logo* (Flamingo: London, 2001), also captured and nurtured a different mood.

88. T. J. Clark, *Farewell to an Idea: Episodes from a History of Modernism* (New Haven: Yale University Press, 1999).

89. "And reassembling our afflicted Powers, / Consult how we may henceforth most offend / Our Enemy, our own loss how repair, / How overcome this dire Calamity, / What reinforcements we may gain from Hope, / If not what resolution from despare." John Milton, *Paradise Lost*, Book 1.

90. Retort (Iain Boal, T. J. Clark, Joseph Matthews, Michael Watts), *Afflicted Powers: Capital and Spectacle in a New Age of War* (London: Verso, 2005), xii.

91. Daniel Bensaïd, preface to the English Edition, *Marx for Our Times*, xi.

92. *Politics-Poetics: Documenta X—the Book*, ed. Catherine David (Ostfilden-Ruit: Hatje Cantz Verlag, 1997).

93. *Documenta 11*, ed. Okwui Enwezor (Ostfildern-Ruit: Hatje Cantz Verlag, 2002). Enwezor gave Retort a platform to exhibit at the Seville Biennial (BIACS2) in Autumn 2006: Okwui Enwezor, ed., *The Unhomely: Phantom Scenes in Global Society* (Barcelona: Actar D, 2006). Despite the criticisms of the curating of Documenta 12 in 2007, in which Robert M. Buergel aimed to avoid the political conception of the previous two Documentas, politicized trends in art were still very visible.

1. T. J. Clark and the Pain of the Unattainable Beyond

1. Arnold Hauser, *The Philosophy of Art History* (London: Routledge and Kegan Paul, 1959), 11–12.

2. T. J. Clark, *Image of the People: Gustave Courbet and the 1848 Revolution* (1973; repr. London: Thames & Hudson, 1982); *The Absolute Bourgeois: Artists and Politics in France 1848–1851* (1973; repr. London: Thames & Hudson, 1982).

3. T. J. Clark, "Jackson Pollock's Abstraction," *Reconstructing Modernism: Art in New York, Paris, and Montreal, 1945–1964*, ed. Serge Guilbaut (Cambridge, MA: MIT Press, 1990), 172–238 (see 239–243 for the discussion from the conference "Hot Paint for Cold War" held at the University of British Columbia in 1986); T. J. Clark, *Farewell to an Idea: Episodes from a History of Modernism* (New Haven: Yale University Press, 1999).

4. T. J. Clark, *The Painting of Modern Life: Paris in the Art of Manet and His Followers* (London: Thames and Hudson, 1984).

5. Meyer Schapiro, "The Nature of Abstract Art" (1936), *Modern Art: 19th and 20th Centuries: Selected Papers* (New York: George Braziller, 1979), 185–211 (originally in *Marxist Quarterly*).

6. For the radical criticisms—some preceding the appearance of "Jackson Pollock's Abstraction"—see, for example, Nicholas Green and Frank

Mort, "Visual Representation and Cultural Politics," *Block* 7 (1982): 59–68; Adrian Rifkin, "Marx' Clarkism," *Art History* 8, no. 4 (1985): 488–495; Frank Mort and Nick Green, "Is There Anyone Here From Education (Again)?: Radical Art and Education for the 1990s," *Block* 12 (1987–1988): 20–7; Griselda Pollock, "Vision, Voice and Power: Feminist Art Histories and Marxism" and "Modernity and the Spaces of Femininity," *Vision and Difference: Femininity, Feminism and the Histories of Art* (London: Routledge, 1988), 18–49 and 50–90. See also Donald Preziosi, *Rethinking Art History: Meditations on a Coy Science* (New Haven: Yale University Press, 1989), 159–168.

7. T. J. Clark, *The Sight of Death: An Experiment in Art Writing* (New Haven: Yale University Press, 2006).

8. In this regard, it is worth noting Peter Osborne's discussion of the character of art criticism (as opposed to art history) as "at once *iterative* and inherently *political*," and specifically his claim that Greenberg's early work and the force of its aesthetic judgment involved "a genuinely 'tensed' conception of historical time." Osborne contrasts this sense of time in art criticism to art history's historicist "progressivism" and to the idealizing, de-temporalizing work of the aesthetic (modes of thinking that he also detects in Greenberg's writing). While Clark mostly addresses art of the past—a perspective more readily associated with the role of the historian—he seizes it in the terms that Osborne claims for contemporary practice and the politics of judgment. See Peter Osborne, "Time and the Artwork," in *Philosophy in Cultural Theory* (London: Routledge, 2000), 82–85; Peter Osborne, "Art Beyond Aesthetics: Philosophical Criticism, Art History and Contemporary Art," *Art History* 27, no. 4 (2004): 651–670.

9. The English Section of the Situationist International (Tim Clarke, Christopher Gray, Charles Radcliffe, Donald Nicholson-Smith), *The Revolution of Modern Art and the Modern Art of Revolution* (1967; repr. London: Chronos Publications, 1994).

10. Francis Mulhern, *Culture/Metaculture* (London: Routledge, 2000).

11. T. J. Clark, "Modernism, Postmodernism, and Steam," *October* 100 (2002): 160.

12. Ibid., 154–174.

13. Retort (Iain Boal, T. J. Clark, Joseph Matthews, Michael Watts), "Afflicted Powers," *New Left Review* 27 (2004): 5–21; *Afflicted Powers* (London: Verso, 2005); T. J. Clark and Donald Nicholson-Smith, "Why Art Can't Kill the Situationist International," *October* 70 (1997): 15–31.

14. T. J. Clark, "Origins of the Present Crisis," *New Left Review* 2 (2000): 85–96.

15. Arnold Hauser, *The Social History of Art*, 4 vols. (1951; repr. London: Routledge and Kegan Paul, 1962).

16. T. J. Clark, "The End of Left Art History?," *Kritische Berichte* 3 (2006): 5–8.

17. T. J. Clark, *Image of the People* (1981), 6 (the same Preface appears in *Absolute Bourgeois*, 6).

18. T. J. Clark, Author's Note, 1984, "Clement Greenberg's Theory of Art," in *Pollock and After: The Critical Debate*, ed. Francis Frascina (London: Harper and Row, 1985), 55.

19. Clark, "Modernism, Postmodernism, and Steam," *October* 100 (2002): 165.

20. T. J. Clark, "A Bourgeois Dance of Death: Max Buchon on Courbet—1," *The Burlington Magazine* April 1969, 208.

21. English Section of SI, *Revolution of Modern Art*, 4.

22. Ibid., 15, 26. The concept of revolutionary praxis as ludic, however, and the argument in favor of delinquency, petty crime, and "the new lumpen" may not meet most of the Left's sense of the revolution and its agent.

23. Timothy J. Clark, "Preliminaries to a Possible Treatment of 'Olympia' in 1865," *Screen* 21, no. 1 (1980): 18–41. An early version of this essay was published in *Histoire et critique des arts*; the final version appeared in *Painting of Modern Life*, 79–146.

24. Clark, "Preliminaries," 39.

25. Ibid., 25, 39, MacCabe cited 37.

26. As one of the few partisans of the position represented by Clark, Jeff Wall not only reworked themes from "the painting of modern life" but also developed an aesthetic that refused the "Godardian" mode then popular among radical practitioners, seeking to exploit, rather than break, the existing strategies of identification. Arielle Pelenc in correspondence with Jeff Wall, *Jeff Wall* (London: Phaidon, 1996), 11. On the strategy of disidentification in art, see Griselda Pollock, "Screening the Seventies: Sexuality and Representation in Feminist Practice—a Brechtian Perspective," *Vision and Difference*, 155–199.

27. Clark, "Preliminaries," 40.

28. Peter Wollen, "Manet: Modernism and Avant-Garde," *Screen* 21, no. 2 (1980): 15–25. For further comment on the Clark/Wollen exchange, Charles Harrison, Michael Baldwin and Mel Ramsden, "Manet's 'Olympia' and Contradiction," *Block* 5 (1981): 34–43.

29. Laura Mulvey, "Changes: Thoughts on Myth, Narrative and Historical Experience," *History Workshop Journal* 23 (1987): 3–19.

30. Serge Guilbaut, "The Relevance of Modernism," *Modernism and Modernity*, ed. Benjamin H. D. Buchloh and Serge Guilbaut (Halifax, Nova Scotia: NSCAD Press, 1983), xi.

31. T. J. Clark, "Clement Greenberg's Theory of Art," *Critical Inquiry* 9, no. 1 (1982): 139–156; Michael Fried, "How Modernism Works: A Response to T. J. Clark," 217–234 in the same issue. This was a special issue based on the conference "The Politics of Interpretation." Clark responded to Fried: Clark, "Arguments About Modernism: A Reply to Michael Fried," in *The Politics of Interpretation*,

ed. W. J. T. Mitchell (Chicago: University of Chicago Press, 1983), 239–248. The debate was republished by Francis Frascina, ed., *Pollock and After: The Critical Debate* (London: Harper and Row, 1985), to which all subsequent references are made.

32. Clark, "Clement Greenberg's Theory of Art," 55.

33. Clark, "Arguments about Modernism," 82.

34. Greenberg always denied that his account had a "logic." See his comments on Mondrian in "Modernist Painting" (1960) and "Complaints of an Art Critic" (1967), in *The Collected Essays and Criticism: Modernism with a Vengeance, 1957–1969*, ed. John O'Brian (Chicago: Chicago University Press, 1993), 85–93 and 265–272.

35. Clark, "Clement Greenberg's Theory of Art," 55.

36. Ibid., 58, 55.

37. Clark, "Arguments about Modernism," 83.

38. Clark, "Clement Greenberg's Theory of Art," 59.

39. Ibid., 52.

40. Clark, "Preface to the New Edition" (1981), in *Absolute Bourgeois*, 7.

41. Clark, "Clement Greenberg's Theory of Art," 59.

42. Clark, "Arguments about Modernism," 82.

43. Charles Harrison, *Essays on Art & Language* (Oxford: Blackwell, 1991), 231.

44. Clark, "Clement Greenberg's Theory of Art," 60.

45. Clark, "Jackson Pollock's Abstraction," 220. Clark counters those who doubt the viability of practices of resistance; such ideas, he argues, are no more "pie in the sky" than the other routine claims about art's "universality," etc. However, in *Farewell*, the same passage indicates greater doubt: "it is hard to tell at present whether ideas of resistance and refusal have any sustaining force still left them, or have been hopelessly incorporated into a general spectacle" (364).

46. Clark, Author's note added in 1984, "Clement Greenberg's Theory of Art," 55.

47. Clark, *Image of the People*, 12.

48. Ibid., 12, 10.

49. Clark, *Absolute Bourgeois*, 57; Clark, "Jackson Pollock's Abstraction," 181.

50. Note Clark's qualms in "Preface to the New Edition" (1981), in *Image of the People*, 6.

51. Michael Podro, *The Critical Historians of Art* (New Haven: Yale University Press, 1982).

52. Hauser, *Philosophy of Art History*, 269, 268, 275.

53. Ibid., 259–261.

54. Antal, "Remarks on the Method of Art History," 188.

55. See, for example, from very different standpoints: Jay Bernstein, "The Death of Sensuous Particulars: Adorno and Abstract Expressionism," *Radical Philosophy* 76 (1996): 7–18; Gilles Deleuze and Félix Guattari, *A Thousand Plateaus: Capitalism and Schizophrenia* (1980; repr. London: Athlone Press, 1988); David Harvey, "Militant Particularism and Global Ambition," *Justice, Nature and the Geography of Difference* (Oxford: Blackwell, 1996), 19–45.

56. Jean-Paul Sartre, *Search for a Method*, trans. Hazel E. Barnes (1960; repr. New York: Vintage Books, 1968), 17n2.

57. Ibid., 53, 37.

58. Ibid., 27–28.

59. Ibid., 126.

60. Cf. Kurt W. Forster, "Critical History of Art, or Transfiguration of Values?" *New Literary History* 3 (1972): 459–470.

61. O. K. Werckmeister, "Marx on Ideology and Art," *New Literary History* 4 (1973): 501–519. See also Werckmeister, "From Marxist to Critical Art History," College Arts Association, session on Marxism and Art History, Chicago (January 1976), unpublished, where the author argues against "ideological generalization" of academic art history and calls for a "critical art history." Even critical art historians are not immune from institutional ideology which accommodates "in an ideological medium both affirmative and critical responses to society" and sustains "the mental disconnection" between contemporary political realities and art history.

62. O. K. Werckmeister, "The Political Ideology of the Bayeux Tapestry," *Studi Medievali*, third series, 17, no. 2 (1976): 589.

63. Clark, *Image of the People*, 17.

64. Clark, "A Bourgeois Dance of Death: Max Buchon on Courbet—2," *The Burlington Magazine* May 1969, 289.

65. Clark, *Image of the People*, 151.

66. Clark, "Preliminary Arguments: Work of Art and Ideology," unpublished paper circulated for the session "Marxism and Art History," College Arts Association, Chicago (January 1976), thesis 1.

67. Clark, "Courbet the Communist and the Temple Bar Magazine," *Block* 4 (1981): 34–35.

68. Clark, *Image of the People*, 158.

69. Clark, "Courbet the Communist," 35, 38.

70. Ibid., 37.

71. Sartre, *Search for a Method*, 126.

72. Clark, *Absolute Bourgeois*, 97.

73. Ibid., 107.

74. Clark, *Image of the People*, 15.

75. See Clark's afterword to "Jackson Pollock's Abstraction," 230.

76. Clark, *Image of the People*, 19.

77. T. J. Clark, "The Conditions of Artistic Creation," *Times Literary Supplement*, 24 May 1974, 561.

78. Ibid., 561, 562.

79. Clark, "Conditions of Artistic Creation," 561.

80. Clark, *Image of the People*, 13.

81. Erwin Panofsky, *Perspective as Symbolic Form*, trans. Gertrud Koch (1927; repr. New York: Zone Books, 1991).

82. Clark, "Jackson Pollock's Abstraction," 181. The citation is from G. W. F. Hegel, *Phenomenology of Spirit*, trans. A. V. Miller (Oxford: Oxford University Press, 1977), 455–456. Cf. Hauser: "No one will wish to assert that an artistic form can be extracted from some extrinsic and alien fact as demonstrably as an egg from a hen" (*Philosophy of Art History*, 266).

83. Clark, "Jackson Pollock's Abstraction," 181.

84. Clark, "In Defense of Abstract Expressionism," *October* 69 (1994): 35; *Farewell*, 388.

85. Clark, "Jackson Pollock's Abstraction," 181.

86. Ibid., 177.

87. Mikhail Bakhtin, "Art and Answerability" (1919), *Art and Answerability: Early Philosophical Essays by M. M. Bakhtin*, trans. Vadim Liapunov (Austin: University of Texas Press, 1990), 1–3. Clark introduces this to his chapter on El Lissitzky to consider the question drawn from Karl Mannheim and posed to art: "who are the protagonists and who the antagonists?" Clark's point here is to interrogate the politics of early Bolshevism under War Communism, especially as they might be seen to intersect with artists' efforts to regain purposefulness for their work; thus, "guilt" and "answerability" take on a more aggressively besieged character. Clark, *Farewell*, 261–263.

88. See the discussion between Clark and Greenberg in Benjamin H. D. Buchloh, Serge Guilbaut, and David Solkin, eds., *Modernism and Modernity*, 192.

89. Clark, "Clement Greenberg's Theory of Art," 59.

90. Harrison, *Essays on Art & Language*, 228.

91. Clark, "Clement Greenberg's Theory of Art," 58.

92. Karl Marx, *Capital: A Critique of Political Economy*, vol. 1, trans. Samuel Moore and Edward Aveling (Moscow: Progress Publishers, 1954), 78, my emphasis.

93. Norman Geras, "Essence and Appearance: Aspects of Fetishism in Marx's *Capital*," *Literature of Revolution: Essays on Marxism* (1971; repr. London: Verso, 1986), 63–84; Peter Dews and Peter Osborne, "The Frankfurt School & the Problem of Critique: A Reply to McCarney," *Radical Philosophy* 45 (1987): 2–11. See also Joseph McCarney, "What Makes Critical Theory 'Critical'?," *Radical Philosophy* 42 (1986): 1–22.

94. Clark, "Clement Greenberg's Theory of Art," 59, my emphasis.

95. Michael Fried, "Three American Painters: Kenneth Noland, Jules Olitski, Frank Stella," in *Art and Objecthood: Essays and Reviews* (1965; repr. Chicago: University of Chicago Press, 1998), 222.

96. Leja refers to popular psychology and philosophy, such as Harvey Ferguson's *Modern Man: His Belief and Behavior* (New York: Alfred A. Knopf, 1936).

97. Michael Leja, "Pollock and Metaphor," *Reframing Abstract Expressionism: Subjectivity and Painting in the 1940s* (New Haven: Yale University Press, 1993), 234.

98. Clark, "Jackson Pollock's Abstraction," 199–200. Note the Adornian gloss, which gives a dialectical recasting, where "Dissonance is the truth about harmony. Harmony is unattainable" (Adorno) and where dissonance is the necessary moment of expression and mimesis (Adorno cited on 200).

99. Cf. the distinction between representing and resembling in Art & Language, "Portrait of V. I. Lenin," *Art-Language* 4, no. 4 (1980): 26–62; Michael Baldwin, Charles Harrison, and Mel Ramsden, "Art History, Art Criticism and Explanation," *Art History* 4, no. 4 (1981): 432–456.

100. Clark, "Jackson Pollock's Abstraction," 199.

101. "Of course in a sense it resisted the metaphors, and the painters we most admire insisted also on it as an awkward, empirical quiddity; but the 'also' is the key word here: there was no fact without the metaphor, no medium without its being the vehicle of a complex act of meaning." (Clark, "Clement Greenberg's Theory of Art," 58.) The "work against metaphor" is used in "Jackson Pollock's Abstraction," 201 (revised to "work is a constant action against metaphor" in *Farewell*, 338).

102. Clark, "Jackson Pollock's Abstraction," 218. Cf. the phrasing in *Farewell*: "Painting now stands or falls exactly by its ability to show what gets in the way of likeness, and to what extremity art must resort if it wants to make likenesses in spite of everything. For modern art . . . has never been driven by a dogmatic wish to avoid the pursuit of resemblance per se, but by the belief that in present circumstances it could only reinvent the possibility of making and matching by having it be exactly that—a possibility, not a foregone conclusion" (*Farewell*, 354).

103. Fried, "Three American Painters," 225.

104. For further discussion, see Gail Day and Chris Riding, "The Critical Terrain of 'High Modernism,'" in *Varieties of Modernism*, ed. Paul Wood (New Haven: Yale University Press, 2004), 189–213.

105. Clark, "Jackson Pollock's Abstraction," 197.

106. Clark, "Jackson Pollock's Abstraction," 197.

107. Ibid., 199.

108. Ibid., 201.

109. Ibid., 221 (Clark drops the capitalizing of "Sign" and "Nature" in *Farewell*, 365).

110. Clark, "Arguments about Modernism," 82.

111. It is seen as a distraction in Malcolm Bull, "Between the Cultures of Capital," *New Left Review* 11 (2001): 95–113.

112. Clark, *Farewell*, 329; "Jackson Pollock's Abstraction," 234n44. The theme of the unhappy consciousness is recurring. See Fredric Jameson, *The Prison House of Language: A Critical Account of Structuralism and Russian Formalism* (Princeton, NJ: Princeton University Press, 1972), 90; Richard Sennett, *Authority* (1980; repr. Faber & Faber, 1993); Donald B. Kuspit, "The Unhappy Consciousness of Modernism," *Artforum* 19 (1981): 53–57. Kuspit cites Greenberg as the key example of this "quasi-Marxist" materialism; "purity is defined as much by what it negates as by what it affirms, and . . . its self-certainty or affirmative character rests on a foundation of uncertainty, a shaky negation" (53–54). Resulting in unhappiness, modernism "is haunted by the desire to communicate beyond itself"; its inward turn to "the language of art" being met by its awareness of "its uselessness for communication; its antidote (literalness and purity) to the insecurities of the present turning, like the *pharmakon*, into an insidious poison, and leading to 'the dead end of presentness'" (55).

113. Hegel, *Phenomenology*, 126. There is debate about the anthropomorphizing of Hegel's terms for a stage of self-consciousness, particularly since Kojève's work and subsequent psychoanalytic readings.

114. Ibid., 126, 137.

115. Fried, "Three American Painters," 227.

116. Clark, "Jackson Pollock's Abstraction," 222.

117. Bakhtin cited by Clark, "Jackson Pollock's Abstraction," 177.

118. The discussion of the "imaginary" public comes from *Image of the People*, 11–12.

119. Ibid., 237n86, and the discussion on 239.

120. Clark, *Image of the People*, 15.

121. T. J. Clark, "All the Things I Said about Duchamp: A Response to Benjamin Buchloh," *The Duchamp Effect*, ed. Martha Buskirk and Mignon Nixon (Cambridge, MA: MIT, 1996), 141–143.

122. See William Empson, "Metaphor," *The Structure of Complex Words* (1951; repr. London: Penguin Books, 1995); Roman Jakobson, *Language in Literature* (Cambridge, MA: The Belknap Press of Harvard University Press, 1987); Hayden White, *Metahistory: the Historical Imagination in Nineteenth-Century Europe* (Baltimore: The Johns Hopkins University Press, 1973), 34.

123. Clark, *Farewell*, 438n40.

124. Fredric Jameson, *The Political Unconscious: Narrative as a Socially Symbolic Act* (London: Cornell University Press, 1981), 100.

125. The distinction of metonymy and synecdoche is much debated. For the synecdochic argument, see Adrian Rifkin, "Marx' Clarkism"; for the metonymic one, see Preziosi, *Rethinking Art History*.

126. Jameson, *Postmodernism*, 238; White, *Metahistory*, 34.

127. See the post-presentation discussion of "Jackson Pollock's Abstraction" in Guilbaut, *Reconstructing Modernism*, 243.

128. Clark, "Jackson Pollock's Abstraction," 181.

129. Clark, "In Defense of Abstract Expressionism," 26.

130. Clark, "Condition of Artistic Creation," 561.

131. Ibid., 562.

132. Hauser, *Philosophy of Art History*, 16, vi.

133. Georg Lukács, *History & Class Consciousness*, 203, 204.

134. Hauser, *Philosophy of Art History*, 12.

135. Lukács, *History & Class Consciousness*, 203.

136. Hegel, *Phenomenology*, 19.

137. T. W. Adorno, *Negative Dialectics*, trans. E.B. Ashton (1966; repr. London: Routledge, 1990), 27–28. Marx Wartofsky also doubts whether Hegel's own thought realized this. Distinguishing the external approach to the dialectic taken by Socrates from the internal/immanent approach of Plato, Marx, and Feuerbach, Wartofsky remarks that "it is not clear to me that Hegel subjects *himself* to an engagement *in* the dialectic, in his own thought." See Marx W. Wartofsky, *Feuerbach* (Cambridge: Cambridge University Press, 1977), 9.

138. Clark, "Modernism, Postmodernism, and Steam," 161–162, 173.

139. Clark, "Jackson Pollock's Abstraction," 201.

140. T. J. Clark, "Freud's Cezanne," *Representations* 52 (1995): 117, 116; *Farewell*, 166, 165

141. Clark, "In Defense of Abstract Expressionism," 22–48.

142. Clark, "End of Left Art History?"

2. Looking the Negative in the Face

1. Italo Calvino, *Invisible Cities*, trans. William Weaver (1972; repr. New York: Harcourt Brace & Company, 1974), 164–165.

2. Office for Metropolitan Architecture (OMA), Rem Koolhaas, and Bruce Mau, *S, M, L, XL*, ed. Jennifer Sigler (Rotterdam: 010 Publishers, 1995), 1200.

3. Ibid., 1208.

4. Ibid.

5. See Paolo Morachiello, "The Department of Architectural History: A Detailed Description," *Architectural Design* 55, no. 516 (1985): 68–71 (1985); and the essays in *Casabella* (January-February 1995), 619–620, a memorial issue to Tafuri.

6. Giulio Lupo and Mercedes Daguerre, "Entrevista con Francesco Dal Co," *Materiales* (Buenos Aires) 5 (1985).

7. Manfredo Tafuri, *Architecture and Utopia: Design and Capitalist Development*, trans. Barbara Luigi La Penta (1969; repr. Cambridge, MA: MIT, 1976). The book was based on "Per una critica dell'ideologia architettonica" for the journal *Contropiano* (1969), translated as "Toward the Critique of Architectural Ideology," trans. Stephen Sartarelli, in *Architecture Theory since 1968*, ed. K. Michael Hays (Cambridge, MA: MIT Press, 1998), 2–35. Manfredo Tafuri, *The Sphere and the Labyrinth: Avant-Gardes and Architecture from Piranesi to the 1970s*, trans. Pellegrino d'Acierno and Robert Connolly (Cambridge, MA: MIT, 1987; Italian edition, 1980). Manfedo Tafuri and Francesco Dal Cò, *Modern Architecture*, 2 vols., trans. Robert Erich Wolf (London: Faber & Faber, 1976; Italian edition, 1976); I draw only on the material that Tafuri prepared.

8. Manfredo Tafuri, ed., *Socialismo, città, architettura URSS 1917–1937: Il contributo degli architetti europei* (Rome: Officina, 1971).

9. Giorgio Ciucci, Francesco Dal Có, Mario Manieri-Elia, and Manfredo Tafuri, *The American City: from the Civil War to the New Deal*, trans. Barbara Luigia La Penta (1973; repr. London: Granada, 1988).

10. For example, Manfredo Tafuri, *Ludovico Quaroni e lo sviluppo dell'architettura moderna in Italia* (Edizione de Comunita, 1964); *Theories and History of Architecture*, trans. Giorgio Verrecchia (1968; repr. London: Granada, 1980); *History of Italian Architecture, 1944–1985*, trans. Jessica Levine (1986; repr. Cambridge, MA: MIT Press, 1989); *Venice and the Renaissance*, trans. Jessica Levine (1985; Cambridge, MA: MIT Press, 1989); *Interpreting the Renaissance: Princes, Cities, Architects* (1992; New Haven: Yale University Press, 2006), the introduction and first chapter of which first appeared in English as "A Search for Paradigms: Project, Truth, Artifice," *Assemblage* 28 (1996: 46–69.

11. Daniel Sherer, "Tafuri's Renaissance: Architecture, Representation, Transgression," *Assemblage* 28 (1996): 42.

12. Joan Ockman, ed., *Architecture, Criticism, Ideology* (Princeton: Princeton Architectural Press, 1985).

13. Mary McLeod, "Introduction," Ockman, ed., *Architecture, Criticism, Ideology*, 7.

14. Ibid., 10.

15. K. Michael Hays, "The Oppositions of Autonomy and History," in *Oppositions Reader* (New York: Princeton Architectural Press, 1998), xiii.

16. Hays, "Oppositions of Autonomy and History," ix.

17. Fredric Jameson, "The Politics of Theory: Ideological Positions in the Postmodernism Debate," *New German Critique* 33 (1984): 53–65.

18. Tafuri, *Theories and History of Architecture*, "Note to the second (Italian) edition" (the note is unpaginated).

19. Jameson, "Politics of Theory," 61.

20. From 1980, Cacciari's approach, while still often focused on Viennese aesthetic and intellectual culture, turns toward a "positive" philosophy and an exploration of "silence." Cacciari's *Architecture and Nihilism: On the Philosophy of Modern Architecture*, trans. Stephen Sartarelli (New Haven: Yale University Press, 1993), brings together a range essays from the nineteen-seventies to the early eighties and a new epilogue, "On the Architecture of Nihilism." Also in English are: "Eupalinos or Architecture," *Oppositions* 21 (1980); *The Necessary Angel*, trans. Miguel E. Vatter (1984; repr. Albany: SUNY, 1994); *Posthumous People: Vienna at the Turning Point*, trans. Rodger Freidman (1980; repr. Stanford: Stanford University Press, 1996).

21. Llorens, "Manfredo Tafuri," 85. Llorens also raises the possibility that Tafuri's analysis is no more than one proceeding by analogy.

22. Ciucci et al., *American City*, xi.

23. Giulio Carlo Argan, "Architecture and Ideology" (1957), in *Architecture Culture 1943–1968: A Documentary Anthology*, ed. Joan Ockman (New York: Rizzoli, 1993), 253–259.

24. Tafuri, *Architecture and Utopia*, 50.

25. The centrality of the technocratic elite in the modern state—here signified by "the Plan"—has been widely commented upon. Notably, for Tafuri, however, it does not seem to be associated with the notion of the end of the working class.

26. Tafuri, *Modern Architecture*, vol. 1, 110.

27. Tafuri, *Architecture and Utopia*, 119.

28. Tafuri, *Modern Architecture*, vol. 1, 110.

29. Georg Simmel, "The Metropolis and Mental Life" (1903), in *On Individuality and Social Forms*, ed. Donald N. Levine (Chicago: University of Chicago Press, 1971), 325.

30. Ibid.

31. Ibid., 326.

32. Ibid.

33. Ibid., 329.

34. Ibid., 329–30.

35. Ibid., 330–1.

36. Massimo Cacciari, "The Dialectics of the Negative and the Metropolis," *Architecture and Nihilism*, 4.

37. Ibid., 4.

38. Ibid., 5–6.

39. Ibid., 6–7. Like many of the Italian debates emerging from *operaismo*, Cacciari, here, does not subsume use value into exchange value.

40. Cacciari, "Dialectic of the Negative," 8. Cf. Cacciari, "Lou's Buttons" (1980) in *Posthumous People.*

41. Tafuri, *Sphere and the Labyrinth*, 119.

42. Tafuri, *Architecture and Utopia*, 78.

43. Guy Debord, *Society of the Spectacle*, trans. Donald Nicholson-Smith (1967; repr. London: Verso, 1994), 143 (thesis 203).

44. Tafuri, *Modern Architecture*, vol. 1, 105.

45. Ibid., 106.

46. The distinction of positive and negative avant-gardes, which Tafuri both works with and unravels, had been disseminated by Renato Poggioli's influential book *The Theory of the Avant-Garde*, trans. Gerald Fitzgerald (1962; repr. Cambridge: Belknap Press, 1968).

47. Tafuri, *Modern Architecture*, vol. 1, 108.

48. Tafuri, *Architecture and Utopia*, 92.

49. Ibid., 95. This inability to combine the elementary forms demonstrated "that no form can be given to the recovery of totality (of being, as of art) except form derived from the problematic nature of form itself."

50. Tafuri, *Modern Architecture*, vol. 1, 112. This is a characteristic move: "decomposition," Tafuri wrote in a discussion of Lu Märten, "showed its constructive side" (*Sphere and the Labyrinth*, 123).

51. Tafuri, *Sphere and the Labyrinth*, 134–5.

52. Tafuri, *Modern Architecture*, vol. 1, 110.

53. Tafuri, "The Uncertainties of Formalism: Victor Shklovsky and the Denuding of Art," *Architectural Design Profile: On the Methodology of Architectural History* (London: Architectural Design, 1981), 74.

54. Shklovsky cited in Tafuri, "Uncertainties of Formalism," 75, 76. Shklovsky's *The Resurrection of the Word* was published in 1914.

55. Tafuri, *Modern Architecture*, vol. 1, 112.

56. Tafuri, *Sphere and the Labyrinth*, 119.

57. Tafuri, *Modern Architecture*, vol. 1, 105.

58. Tafuri, "Uncertainties of Formalism," 77.

59. Ibid.

60. Tafuri, *Sphere and the Labyrinth*, 96.

61. Ibid., 110.

62. Tafuri, *Modern Architecture*, vol. 1, 112.

63. Tafuri, *Sphere and the Labyrinth*, 145.

64. Ibid., 17, 19.

65. Ibid., 137, my emphasis.

66. Tafuri, *Modern Architecture*, vol. 1, 105. With "loss of a center," Tafuri alludes not only to modernity's general crisis of legitimation but also to a specific postwar publication, *Verlust der Mitte—Die bildende Kunst des 19. und 20. Jahrhunderts als Symptom und Symbol der Zeit*, by Right-wing art historian Hans Sedlmayr. Tafuri's oeuvre can be seen as a risposte to Sedlmayr's interpretation of themes addressed in his book (inter alia: autonomy, the demise of the composite "work of art," dehumanization, fragmentation, and the "negation of art").

67. Tafuri, *Sphere and the Labyrinth*, 131.
68. Ibid., 100; *Modern Architecture*, vol. 1, 106.
69. Tafuri, *Architecture and Utopia*, 89.
70. Tafuri, *Sphere and the Labyrinth*, 105.
71. Tafuri, *Architecture and Utopia*, 134–135.
72. Ibid., 56. Mannheim, Keynes, and Weber, Tafuri argues, articulated this new dynamic conception.
73. Tafuri, *Theories and History*, 63.
74. Tafuri, *Architecture and Utopia*, 96.
75. Ibid., 62, 61.
76. Tafuri, *Theories and Histories*, 31.
77. Tafuri, *Architecture and Utopia*, 56, my emphases.
78. Tafuri, *Modern Architecture*, vol. 1, 110; *Theories and History*, 36.
79. Tafuri, "The Disenchanted Mountain," in *American City*, 493–503.
80. Ibid., 503. There are clear parallels between Tafuri's analysis of hypermodern skyscrapers and Jameson's account of John Portman's Bonaventura Hotel in Los Angeles (which he was developing while preparing essays on, or involving, Tafuri). See Fredric Jameson, "Postmodernism, or, the Cultural Logic of Late Capitalism," *New Left Review* 146 (1984): 80–84. Jameson's main essay on Tafuri was presented at the Revisions's symposium in 1982, and published three year's later: Fredric Jameson, "Architecture and the Critique of Ideology," *Architecture, Criticism, Ideology*, ed. Ockman (Princeton: Princeton Architectural Press, 1985), 51–87. Tafuri's work also plays an important role in Jameson's "Politics of Theory" (1984). While altering the political argument, a Tafurian analytical framework shapes a number of discussions to be found in Jameson's "Postmodernism" (from the Bonaventura discussion to Jameson's rejection of nostalgia and alienation). The debt owed to Tafuri and Cacciari, specifically their concept of temporality, is expressed in passing in Fredric Jameson, *Archaeologies of the Future: The Desire Called Utopia and Other Science Fictions* (London: Verso, 2005), 228.
81. Tafuri, *Architecture and Utopia*, 178.
82. Ibid., 100.
83. Ibid., ix
84. Ibid., 145–148.
85. Tafuri, "There Is No Criticism, Only History," *Design Book Review* 9 (1986): 11.
86. Tafuri, *Sphere and the Labyrinth*, 148.
87. Tafuri, *Architecture and Utopia*, 148; *Sphere and the Labyrinth*, 148.
88. Tafuri, *Architecture and Utopia*, 148.
89. Tafuri, *Sphere and the Labyrinth*, 112.
90. T. J. Clark, "Jackson Pollock's Abstraction," *Reconstructing Modernism: Art in New York, Paris, and Montreal 1945–1964*, ed. Serge Guilbaut (Cambridge, MA: MIT Press, 1986), 178. The essay was slightly revised and

retitled as "The Unhappy Consciousness" in *Farewell to an Idea: Episodes from a History of Modernism* (New Haven: Yale University Press, 1999). One of Clark's revisions was to replace "rank pessimists" with "members of modernism's official (late) opposition"; he retains "sardonic glumness" to describe Tafuri (306).

91. Jameson, "Architecture and the Critique of Ideology," 58, 87, 72.

92. Jameson, "Politics of Theory," 60.

93. This point has also been understood by Anthony Vidler in a paper circulated for the conference "Marx, Architecture and Modernity," University of Westminster, convened by the Polytechnic Research Group (School of Architecture) and the English Colloquia, May 2004.

94. Hélène Lipstadt and Harvey Mendelsohn, "Philosophy, History, and Autobiography: Manfredo Tafuri and the 'Unsurpassed Lesson' of Le Corbusier," *Assemblage* 22 (1994): 59.

95. Hays, "Oppositions of Autonomy and History," xiv.

96. The historic compromise was formulated in the early seventies by Enrico Berlinguer, secretary of the PCI from 1972, and advocated a social and political alliance between the Communists, Socialists, and Christian Democrats. Initiated as a defensive strategy against the dangers of reaction (in Italy itself and also with the recent military coup in Allende's Chile in mind), it claimed to pursue the legacy of the antifascist alliances of the nineteen-forties. It paralleled Christian with Communist morality around ideas of collectivist austerity and thought that stability could be achieved by avoiding polarity between Left and Right. The historic compromise—which is now seen as a milestone in the decomposition of the PCI—is widely believed to have been a disaster, which seriously underestimated the deep conservatism of the Christian Democrats and which simultaneously misjudged the dominant morality among the working class.

97. Ivan Turgenev, *Fathers and Sons*, trans. Rosemary Edmonds (1862; repr. London: Penguin, 1975), 94.

98. Tafuri, *Sphere and the Labyrinth*, 148. Tafuri specifically cites the manifesto of the International Union of Neoplastic Constructivists of 1922: "This international is not the result of some humanitarian, idealistic, or political sentiment, but springs from the same amoral and elementary principles on which both science and technology are based" (cited 147–148).

99. Cacciari, "Dialectic of the Negative," 10, 21, 20.

100. Tafuri, *History of Italian Architecture*, 198, 199.

101. Tafuri, *Architecture and Utopia*, 74.

102. Tafuri, *Sphere and the Labyrinth*, 135; *Modern Architecture*, vol. 1, 108, my emphasis.

103. Tafuri, *Sphere and the Labyrinth*, 119.

104. Ibid., 1–21; see also "There Is No Criticism," 8–11.

105. Tafuri, *Sphere and the Labyrinth*, 13, 148.

106. Tafuri, *Architecture and Utopia*, 76.

107. See, for example, Giulio Carlo Argan, *The Baroque Age* (New York: Rizzoli, 1989); originally published as part of *Europe of the Capitals, 1600–1700* (Geneva: Skira, 1964).

108. Tafuri, *Architecture and Utopia*, 21. The conception deployed by both Argan and Tafuri is a direct engagement with the categories of Heinrich Wölfflin's *Principles of Art History*.

109. Llorens, "Manfredo Tafuri," 94n28.

110. Jürgen Habermas, *The Philosophical Discourse of Modernity: Twelve Lectures* (Cambridge: Polity, 1987); Roy Bhaskar, *Dialectic: The Pulse of Freedom* (London: Verso, 1993).

111. Karl Löwith, "The Historical Background of European Nihilism" (1943), in *Nature, History, and Existentialism: and Other Essays in the Philosophy of History* (Evanston: Northwestern University Press, 1966), 3–16; Otto Pöggeler, "Hegel und die Anfänge der Nihilismus-Diskussion," *Man and World* 3 (1970): 163–199; David Breazeale, "The Hegel-Nietzsche Problem," *Nietzsche-Studien* 4 (1975): 146–164; Giorgio Agamben, *Language and Death: The Place of Negativity* , trans. Karen E. Pinkus and Michael Hardt (1982; repr. Minneapolis: University of Minnesota Press, 1991).

112. Cacciari, "The Dialectics of the Negative," 9.

113. Ibid., 16.

114. Ibid., 20.

115. Friedrich Nietzsche, *The Will to Power*, trans. Walter Kaufmann and R. J. Hollingdale (New York: Vintage Books, 1968), §2 and §28, see also Preface.

116. Ibid., §585.

117. Ibid., §32.

118. Gianni Vattimo, *The End of Modernity: Nihilism and Hermeneutics in Post-modern Culture*, trans. Jon R. Snyder (1985; repr. Cambridge: Polity, 1988).

119. Tafuri, *History of Italian Architecture*, 198–199.

120. Ibid., 199. Cf. Tafuri, "The Historical Project," *Sphere and the Labyrinth*, 11; "The Ashes of Jefferson," *Sphere and the Labyrinth*, 291–303; and "There Is No Criticism," 8–11. Elsewhere, Tafuri dismisses the "crisis of modernism" as "merely fashionable, social chit-chat." Doubt has a long history, he argued, and "our century's most attentive and perhaps tragic forms of cultural awareness have always insisted that we work our way through this crisis, rather than lulling ourselves to sleep with it"; postmodern ambiguity serves to reassure us concerning all that is disquieting about capitalism. Tafuri in interview with Giacinto Di Pietrantonio, *Flash Art International* 145 (1989): 67, 68.

121. Cacciari, "Epilogue: On the Architecture of Nihilism," *Architecture and Nihilism*, 209.

122. Tafuri, *History of Italian Architecture*, 199, 191.

123. Ibid., 199. Recently, scholarly distinctions between "modern" and "postmodern" interpretations of nihilism do not capture the position presented by Caciari and Tafuri.

124. Cacciari, "On the Architecture of Nihilism," 209.

125. G. W. F. Hegel, *The Phenomenology of Spirit*, trans. A. V. Miller (Oxford: Oxford University Press, 1977), 19. Gillian Rose compared Nietzsche's self-completing nihilism (*sich vollendenden Nihilismus*)—which she translated as "self-perficient"—with Hegel's *sich vollbringende Skeptizismus*. Gillian Rose, *Dialectic of Nihilism: Poststructuralism and Law* (Oxford: Basil Blackwell, 1984).

126. Tafuri, *Sphere and the Labyrinth*, 106.

127. Tafuri, *Modern Architecture*, vol. 1, 108, my emphasis.

128. Gilles Deleuze, *Nietzsche and Philosophy*, trans. Hugh Tomlinson (1962; repr. London: The Athlone Press, 1983).

129. Tafuri, *Sphere and the Labyrinth*, 106.

130. Cacciari, "Dialectics of the Negative," 19.

131. Ibid., 20.

132. Tafuri, *Sphere and the Labyrinth*, 13. Cacciari would later put more emphasis on listening. See Cacciari, "On the Architecture of Nihilism."

133. Tafuri, "There Is No Criticism," 10, 11. Cf. Tafuri's observation of Shklovsky: "criticism takes over the instruments hitherto belonging to the object of its analysis" (Tafuri, "Uncertainties of Formalism," 76).

134. Tafuri, *Architecture and Utopia*, vii–viii. Cf. comments by Tafuri et al.: certain "critical formulas," they noted, "are simply taken for granted," and readers should be able to detect these simply from the research focus and the reconstruction of historical material (in *American City*, xi).

135. Published quarterly from 1968–1971 (Florence: La Nuova Italia), the journal was also the forum for essays by Dal Cò, Rita di Leo, and Asor Rosa.

136. Mario Tronti, *Operai e Capitale* (Turin: Einaudi, 1966).

137. Cacciari, "Entretien Marco Biraghi avec Massimo Cacciari."

138. On the Italian Left and its context: Paul Ginsborg, *A History of Contemporary Italy: Society and Politics 1943–1988* (London: Penguin, 1990); Robert Lumley, *States of Emergency: Cultures of Revolt in Italy from 1968 to 1978* (London: Verso, 1990); Steve Wright, *Storming Heaven: Class Composition and Struggle in Italian Autonomist Marxism* (London: Pluto, 2002); John Merrington, "Introduction to Negri's 'Keynes and the Capitalist Theory of the State post-1929'" in Toni Negri, *Revolution Retrieved: Selected Writings on Marx, Keynes, Capitalist Crisis and New Social Subjects (1967–83)*, trans. P. Saunders and E. Bostanjoglu (London: Red Notes, 1988); Chris Harman, *The Fire Last Time:*

1968 and After (London: Bookmarks, 1988); Tobias Abse, "Judging the PCI," *New Left Review* 153 (1985), 5–43. On his fellowship in the United States in 1959, Calvino tried unsuccessfully to make contact with Paul Mattick on behalf of Raniero Panzieri, cofounder and chief editor of Quaderni Rossi and the political editor at Calvino's publisher, Einaudi. Italo Calvino, *Hermit in Paris: Autobiographical Writings*, trans. Martin McLauglin (London: Jonathan Cape, 2003), 50.

139. Raniero Panzieri, "The Capitalist Use of Machinery: Marx Versus the 'Objectivists'" (1961), in *Outlines of a Critique of Technology*, ed. Phil Slater (London: Ink Links, 1980), 44–68.

140. Mario Tronti, "Workers and Capital" (1971), *Telos* 14 (1972): 25–62; 58–60 for the discussion of massification. By 1970, PO had collapsed the distinction that they had hitherto made between manual and technical workers, replacing it with a single notion of labor (Wright, *Storming Heaven*, 106).

141. Mario Tronti, "Social Capital" (1971), *Telos* 17 (1973): 116–117.

142. Ginsborg, *History of Contemporary Italy*, 298, 313.

143. See Pier Vittorio Aureli, *The Project of Autonomy: Politics and Architecture within and against Capitalism* (New York: Buell Center and Princeton Architectural Press, 2008)

144. Franco Piperno (1969) and Alberto Asor Rosa (1967), cited in Wright, *Storming Heaven*, 117, 75. According to Wright, "tactical entrism" and "autonomy" were each solutions to the same set of problems. There is some debate, however, whether "entrism" is the correct term to use as the intellectuals who made this move into the PCI appear not to have organized themselves so consciously. I would like to thank Matteo Mandarini, Peter Thomas, Max Tomba, and Alberto Toscano for sharing their perspectives on this moment.

145. Wright, *Storming Heaven*, 99.

146. Asor Rosa's work is cited prominently in both *Architecture and Utopia* and *Theories and History of Architecture*. In the latter, Tafuri attributes his "analysis of the 'active' role of negation," not only to Cacciari's negative thought but also to Asor Rosa's essay "L'uomo, il poeta," from *Angelus Novus* 5–6 (1965), 1–30.

147. Antonio Negri in conversation with Anne Dufourmantelle, *Negri on Negri*, trans. M.B. DeBevoise (New York: Routledge, 2004), 168.

148. Cesco Chinello, "Il Sessanotto operaio e studentesco a Porto Marghera," *Sindicato e lotte dei lavoratori a Padova e nel Venezia (1945–1969)* (Padua: Centro Studi Ettore Luccini, 1998).

149. Lumley, *States of Emergency*, 88. The reference is probably to Rosaldo Bonicalzi.

150. Tafuri, *Architecture and Utopia*, 173.

151. Panzieri, "Capitalist Use of Machinery," 44–68. Panzieri draws especially on chapters 13 through 15 of Marx's *Capital*, vol. 1. Panzieri's arguments

were further developed in "Surplus Value and Planning" (1964), in *The Labour Process and Class Strategies* (London: Conference of Socialist Economists, 1976).

152. Tronti, "Workers and Capital," 27.

153. Ibid., 60

154. Ibid., 53. Paul Piccone's introduction to this translation of "Workers and Capital" was particularly damning of this evaluation.

155. Tronti, "Social Capital," 109, 111.

156. See for example *Potere Operaio*: "Contro il piano" (1967, no. 1); "La classe operaia contro il piano dei padroni" (1967, no. 3); "Classe Operaia e Piano in Emilia" (1967, no. 5).

157. POv-e cited Wright, *Storming Heaven*, 97–98.

158. See Aureli, *Project of Autonomy*.

159. Panzieri, "Capitalist Use of Machinery," 49.

160. Ibid., 57. Panzieri treats Soviet planning under Stalin in much the same manner as he would capitalist planning (and Tafuri followed his argument). However, Panzieri insisted, "it is important not to lose sight of the distinctive feature of Soviet planning compared with the capitalist plan: the authoritative, despotic element of productive organization *arises* within capitalist relations, *survives* in planned economies of a bureaucratic type" (ibid., 57n30). Thus he contrasted the "organic" character of capitalism's "despotism" with its "precarious" quality in the USSR and the People's Democracies.

161. Lumley, *States of Emergency*, 64.

162. Ginsborg, *History of Contemporary Italy*, 292.

163. Panzieri, "Capitalist Use of Machinery," 52.

164. See Tafuri, *Modern Architecture 1*; *Sphere and the Labyrinth*; "'Machine et mémoire': The City in the Work of Le Corbusier," *Urbanisme, Algiers and Other Buildings and Projects, 1930–1933* (Garland Publishing and Fondation Le Corbusier), xxxi–xlvi.

165. For example, Manfredo Tafuri, "Austromarxismo e città: 'Das rote Wein,'" *Contropiano* 2 (1971), 259–311; *Vienna rossa: la politica residenziale nella Vienna socialista, 1919–1933* (Milan: Electa, 1980). The discussion is also important in a number of chapters in *Modern Architecture 1*, especially "The Attempts at Urban Reform in Europe Between the Wars". See also "Sozialpolitik and the City in Weimer Germany," in *Sphere and the Labyrinth*, 197–233.

166. On Tafuri and Tronti's different political assessments of Austro-Marxism and Red Vienna, see Aureli, *Project of Autonomy*, 49–53.

167. Joan Ockman has described how, on account of his PCI membership, Tafuri's visit to the United States in 1974 was restricted to a three-day visa. Joan Ockman, "Venice and New York," *Casabella* January-February 1995: 619–620n4.

168. Lumley, *States of Emergency*, 129.

169. Toni Negri, "Keynes and the Capitalist Theory of the State post-1929," *Revolution Retrieved: Selected Writing on Marx, Keynes, Capitalist Crisis and New Social Subjects (1967–83)*, trans. P. Saunders and E. Bostanjoglu (London: Red Notes, 1988).

170. Negri, "Keynes," 11.

171. Tronti, "Workers and Capital," 44.

172. For Negri, the post-1929 crisis is not primarily significant for initiating recourse to state intervention (which had, he tells us, been a growing tendency since 1871); nor is it characterized by a shift from "liberal" to "totalitarian" state (which, he says, confuses the regime with the state form).

173. Negri, "Keynes," 13, 28–29.

174. Ibid., 13.

175. Ibid., 13, 28.

176. Ibid., 13.

177. Mario Tronti, "Lenin in England" (1964), in *Working Class Autonomy and the Crisis: Italian Marxist Texts of the Theory and Practice of a Class Movement: 1964–79* (London: Red Notes, 1979), 3, 5.

178. Tronti, "Workers and Capital," 46.

179. Tafuri cites 1929 as the "turning point in the history of the cities in the United States," not because of the degree of public control initiated under the New Deal, but because the scale of planning—the "source of capital accumulation" and "place of autonomous production"—escalated from city to region. (*American City*, 432).

180. Negri, "Keynes," 13, 14, 34.

181. Negri, "Labor in the Constitution," in *Labor of Dionysus: A Critique of the State-Form*, Michael Hardt and Antonio Negri (Minneapolis: University of Minnesota Press, 1994), 135.

182. Tronti, "Social Capital," 121.

183. Tronti, "Lenin in England," 1.

184. Negri, "Keynes," 12.

185. Negri, *Revolution Retrieved*, 13.

186. Tronti, "Lenin in England," 1.

187. Negri, "Labor in the Constitution," 59; and "Keynes," 28.

188. Tafuri, *Architecture and Utopia*, 61.

189. Negri, "Keynes," 13.

190. Tronti, "Lenin in England," 2, 6, 5.

191. Negri, "Labor in the Constitution," 61.

192. Tronti, "Social Capital," 121; see also Tronti's "The Strategy of Refusal," *Working Class Autonomy and the Crisis: Italian Marxist Texts of the Theory and Practice of a Class Movement: 1964–79* (London: Red Notes, 1979), 7–21.

193. Tronti, "Lenin in England," 3–4.

194. Cacciari, "The Dialectics of the Negative," 16.

195. Tronti, "Lenin in England," 1.

196. Paul Piccone, "Introduction to Tronti's 'Workers and Capital'," *Telos* 14 (1972). Piccone was, at this point, still committed to salvaging the American New Left. See also Paul Piccone, "Introduction to Bologna's 'Class Composition and Theory of the Party,'" *Telos* 13 (1972): 1–3.

197. Paolo Virno, *A Grammar of the Multitude: For an Analysis of Contemporary Forms of Life*, trans. Isabella Bertoletti, James Cascaito, and Andrea Casson (Los Angeles: Semiotext(e), 2004). Cf. Negri, *Negri on Negri*; Hardt and Negri, *Empire* (Harvard University Press, 2000); Paolo Virno and Michael Hardt, eds., *Radical Thought in Italy: A Potential Politics* (Minneapolis: University of Minnesota Press, 1996); Michael Hardt and Antonio Negri, *Labor of Dionysus*.

198. See comments on Trotsky's "class point of view" in *American City*, xi.

199. Cf. Tafuri, "Note to the second (Italian) edition," *Theories and History*, note unpaginated.

200. Tafuri interview with Di Pietrantonio, 68. Tafuri argued that the avant-garde had not been defeated: "what you have is realization, rather than failure" (68).

201. Fredric Jameson, *A Singular Modernity: Essay on the Ontology of the Present* (London: Verso, 2002).

202. Tafuri, *Theories and History*, 1–2.

203. Tafuri, "Note to the second (Italian) edition," *Theories and History*, note unpaginated.

3. Absolute Dialectical Unrest

1. G. W. F. Hegel, *Phenomenology of Spirit*, trans. A. V. Miller (Oxford: Oxford University Press, 1977), 472.

2. Craig Owens, "The Allegorical Impulse: Towards a Theory of Postmodernism," *October* 12 (1980): 70.

3. *October* was founded in Spring 1976 by two former editors of *Artforum*: Krauss and Annette Michelson; Jeremy Gilbert-Rolfe joined them as the third founding editor. In her analysis of *Artforum*, Amy Newman provides a detailed account of the art-critical background of *Artforum*'s transition from formalism to antiformalism. Amy Newman, *Challenging Art: Artforum 1962–1974* (New York: Soho Press, 2000).

4. Edwin Honig, *Dark Conceit: The Making of Allegory* (1959; repr. New York: Oxford University Press, 1966); Angus Fletcher, *Allegory. The Theory of a Symbolic Mode* (Ithaca: Cornell University Press, 1964); Gay Clifford, *The*

Transformations of Allegory (London: Routledge and Kegan Paul, 1974); Maureen Quilligan, *The Language of Allegory: Defining the Genre* (Ithaca: Cornell University Press, 1979). Although there are other writings that address specific historical examples of allegory, these authors are distinctive in seeking to define the wider historical dimensions and fortunes of allegory as a literary mode. Paul de Man's work, which was to become especially central for the debates of the nineteen-eighties, also needs to be mentioned in this context. De Man took a more theoretical and deconstructive approach: Paul de Man, "The Rhetoric of Temporality" (1969) in Charles Singleton, ed., *Interpretation* (Baltimore: Johns Hopkins University Press, 1969); republished in *Blindness and Insight: Essays in the Rhetoric of Contemporary Criticism* (London: Methuen, 1983); all citations are to the latter. See also Paul de Man, *Allegories of Reading: Figural Language in Rousseau, Nietzsche, Rilke, and Proust* (New Haven: Yale University Press, 1979).

5. Quilligan, *Language of Allegory*, 155.

6. Ibid., 216, 203n67.

7. Rosalind Krauss, "Grids," *October* 9 (1979): 9. Although Greenberg downgraded narrative (along with the haptic) in his analysis of painting in favor of its "optical" qualities, he nevertheless identified (and approved of) it as the central feature of photography. On the optical, see especially Clement Greenberg, "Sculpture in Our Time" (1958) and "Modernist Painting" (1960), *The Collected Essays and Criticism*, in 4 volumes, ed. John O'Brian (Chicago: University of Chicago Press, 1993 [the first 2 volumes appeared in 1986]), vol. 4: 55–61 and vol. 4: 85–93. Fried discusses the different emphases, or "double valence," in Greenberg's use of opticality in Michael Fried, "An Introduction to My Art Criticism," *Art and Objecthood: Essays and Reviews* (Chicago: University of Chicago Press, 1998), 19–23. As Fried notes, "Nothing . . . has come in for more sustained assault in recent years than the claim that modernist painting posits or privileges or establishes the illusion of a purely visual or 'optical' space, one addressed to eyesight alone." On photography, see Clement Greenberg, "Four Photographers" (1964), *The Collected Essays and Criticism*, vol. 4, ed. John O'Brian (Chicago: University of Chicago Press, 1993), 183–187.

8. Craig Owens, "Earthwords," *October* 10 (1979): 126. For a resume of the *October* project, giving particular attention to its interest in French Critical Theory, see Peter Muir, "*October*: La Glace sans tain," *Cultural Values* 6, no. 4 (2002): 419–441.

9. Crimp joined *October* as an editorial associate in Autumn 1977 (vol. 4), progressing through the roles of managing editor and executive editor to become editor in Spring 1986 (vol. 36); he last held this position in Winter 1989 (vol. 51) and his resignation letter appears in vol. 53. Owens was associate editor for five issues from Autumn 1979 (vol. 10). A participant in an interview with Krauss and Michelson in Spring 1980 (vol. 12), Buchloh published regularly in

the journal from Spring 1981 (vol. 16), joining the board in advisory capacity in Winter 1986 (vol. 39) and becoming editor in Spring 1991 (vol. 56).

10. Krauss cited in Muir, "*October*," 421.

11. Krauss cited in Muir, "*October*," 419.

12. Crimp cited in Muir, "*October*," 420.

13. Craig Owens, "The Allegorical Impulse: Towards a Theory of Postmodernism," *October* 12 (1980): 67–86, and "The Allegorical Impulse: Towards a Theory of Postmodernism Part 2," *October* 13 (1980): 58–80. Also published in *October* were Joel Finemann, "The Structure of Allegorical Desire," *October* 12 (1980): 48–66, and Stephen Melville, "Notes on the Reemergence of Allegory, the Forgetting of Modernism, the Necessity of Rhetoric, and the Conditions of Publicity in Art and Criticism," *October* 19 (1981): 55–92. Other articles addressed to allegory at this time include Joan Simon, "Double Takes," *Art in America*, October 1980, and Benjamin H. D. Buchloh, "Allegorical Procedures: Appropriation and Montage in Contemporary Art," *Artforum* 21, no. 1 (1982): 43–56. Hal Foster picks up the theme indirectly—and by way of Buchloh's essay especially—in "Subversive Signs," *Art in America*, November 1982, republished in Hal Foster, *Recodings: Art, Spectacle, Cultural Politics* (Seattle: Bay Press, 1985), 99–117.

14. Owens, "Allegorical Impulse," vol.1, 70.

15. Ibid., vol. 2, 70; vol. 1, 81–82; vol. 2, 79–80.

16. Ibid., vol. 2, 80.

17. Ibid., vol. 1, 71.

18. Rosalind Krauss, "Video: The Aesthetics of Narcissism," *October* 1 (1976): 50–64; "Notes on the Index: Seventies Art in America," *October* 3 (1977): 68–81 and 4 (1977): 58–67; "Stieglitz/*Equivalents*," *October* 11 (1979): 129–140; "The Photographic Conditions of Surrealism," *October* 19 (1981): 3–34; "Nightwalkers," *Art Journal* 41, no. 1 (1981): 33–38; "Photography's Discursive Spaces: Landscape/View," *Art Journal* 42, no. 4 (1982): 311–319. A number of these essays are republished in an especially influential volume: Krauss, *The Originality of the Avant-Garde and Other Modernist Myths* (Cambridge, MA: MIT, 1986).

19. Craig Owens, "Photography *en abyme*," *October* 5 (1978): 73–88; citation from 77.

20. Ibid., 85–86; my emphasis.

21. Douglas Crimp, "Positive/Negative: a Note on Degas's Photographs," *October* 5 (1978): 89–100.

22. Douglas Crimp, "On the Museum's Ruins," *October* 13 (1980): 53; revised in Crimp, *On the Museum's Ruins* (Cambridge, MA: MIT, 1993), 56.

23. Douglas Crimp, "The Museum's Old, the Library's New Subject" (1981), *Museum's Ruins*, 77.

24. Douglas Crimp, "The Photographic Activity of Postmodernism," *October* 15 (1980): 98; revised for *Museum's Ruins*, 117, where "and" becomes "or".

25. Buchloh, "Allegorical Procedures," 43.

26. Ibid., 44. Accordingly, Greenberg's resistance to Duchamp, despite its grounding of "conservative formalism," probably contained nevertheless "a moment of radical truth" (47). Here Buchloh gives an Adornian gloss to the subject. Buchloh refers to Walter Benjamin's late essay, "Central Park" (1939–1940).

27. Ibid., 46–47.

28. Ibid., 47.

29. Ibid., 48.

30. Ibid., 47.

31. Ibid., 48.

32. Douglas Crimp, "Pictures," *October* 8 (1979): 74.

33. Foster, "Subversive Signs," 100.

34. Ibid., 108.

35. Ibid., 103.

36. Owens, "Allegorical Impulse," vol. 1, 64; vol. 2, 79.

37. Ibid., vol. 1, 83.

38. Walter Benjamin, *The Origin of German Tragic Drama*, trans. John Osborne (London: New Left Books, 1977), 159; all reference taken from this edition, hereafter referred to as *Trauerspiel*. Written between 1924 and 1925, Benjamin's *Ursprung des deutschen Trauerspiels* originally appeared in 1928, and then in 1955 in *Schriften*, reissued in 1963 (Frankfurt: Suhrkamp, 1963), and again, in 1974 in the first volume of *Gesammelte Schriften*, ed. Rolf Tiedemann and Hermann Schweppenhäuser. Other texts by Benjamin are also important to the debates of allegory, especially: Walter Benjamin, "Paris—Capital of the Nineteenth Century" (written 1935, published *GS*), trans. Quintin Hoare, *New Left Review* 48 (1968); republished as "Paris, the Capital of the Nineteenth Century," *Selected Writings, vol. 3, 1935–1938*, trans. Howard Eiland (Cambridge, MA: The Belknap Press of Harvard University Press, 2002), 32–49; "Central Park" (written 1938–1939 and published in *GS*), *Selected Writings, vol. 4, 1938–1940*, trans. Edmund Jephcott and Howard Eiland (Cambridge, MA: The Belknap Press of Harvard University Press, 2003), 160–199. Subsequent references for these essays are to *Selected Writings*.

Anglophone attention to Benjaminian allegory largely follows on the heels of the 1977 English translation of the *Trauerspiel*. (Engagement with Benjamin is noticeably absent, or cursory, in the literary studies cited earlier.) See, for example, Terry Eagleton, *Walter Benjamin, or Towards a Revolutionary Criticism* (London: Verso, 1981); Bainard Cowan, "Walter Benjamin's Theory of Allegory," *New German Critique* 26 (1982): 109–122. An important exception is Fredric Jameson, *Marxism and Form: Twentieth-century Dialectical Theories of Literature* (Princeton: Princeton University Press, 1971). Jameson discusses the place of allegory in the thought of both Benjamin and Ernst Bloch, and develops a counterintuitive argument that links together Marxist and Lukácsian concepts of literary realism—and especially the idea of "typification"—with the allegorical form. Benjamin's conception of allegory and montage was subject to considerable attention in Germany in the nineteen-seventies and provides the context for Buchloh's engagement. See, for example: Peter Bürger, *Theorie der Avantgarde* (Frankfurt: Suhrkamp, 1974); Ansgar Hillach, "Allegorie, Bildraum, Montage," *Theorie der Avantgarde* (Frankfurt: Edition Suhrkamp, 1976), 105–142; Harald Steinhagen, "Zu Walter Benjamin's Begriff der Allegorie," *Form und Funktionen der Allegorie* (Stuttgart: Metzler, 1979); Jürgen Naeher, *Walter Benjamin's Allegorie-Begriff als Modell* (Frankfurt: Klett-Cotta, 1975).

39. Benjamin, *Trauerspiel*, 160–161.

40. Honig, *Dark Conceit*, 50. See also Andrew Bowie, *Aesthetics and Subjectivity from Kant to Nietzsche* (Manchester: Manchester University Press, 1990). Bowie argues that, along with its critique of Western metaphysics and representation, Romantic philosophy provided the precedent for the "linguistic turn" of the twentieth century (22).

41. In particular Benjamin charts the debate through the writings of German art historians, philosophers, and aestheticians. As his targets he has Friedrich Creuzer and Joseph von Görres from the early nineteenth century; Hermann Cohen and Carl Horst (both Neo-Kantians); and, from the early twentieth century, Karl Giehlow (who he treats more favorably).

42. Bloch's distinction draws on Benjamin: allegory is "On The Way." See Ernst Bloch, *The Principle of Hope*, vol. 1, trans. Neville Plaice, Stephen Plaice and Paul Knight (1938–1947, revised 1953 and 1959; Cambridge, MA: MIT Press, 1995), 175–176. The discussion of allegory and symbol falls in Part Two, "Anticipatory Consciousness," in the subsection entitled "Discovery of the Not-Yet-Conscious or of Forward Dawning."

43. Benjamin, *Trauerspiel*, 160.

44. De Man, "Rhetoric of Temporality," 189, 207.

45. Samuel Taylor Coleridge, *The Statesman's Manual; or, the Bible the Best Guide to Political Skill and Foresight: a Lay Sermon, Addressed to the Higher Classes of Society, With an Appendix, Containing Comments and Essays Connected with the Study of the Inspired Writings* (London: Gale & Fenner, J. M. Richardson, Hatchard, 1879), 36.

46. Clement Greenberg, "Recentness of Sculpture" (1967), *Collected Essays*, vol. 4, 253–254.

47. In Fried the term "presence" conveys not the aesthetic fullness and immediacy of the symbol (he uses the word "presentness" in this sense), but simply a quotidian human presence. Michael Fried, "Art and Objecthood" (1967), *Art and Objecthood*, 148–172.

48. Benjamin, *Trauerspiel*, 165.

49. Honig, *Dark Conceit*, 114.

50. Ibid., 12.

51. Quilligan, *Language of Allegory*, 29.

52. Fletcher, *Allegory*, 96.

53. Honig, *Dark Conceit*, 4, 5.

54. De Man, "Rhetoric of Temporality," 209.

55. Benjamin, *Trauerspiel*, 208. Owens uses almost the identical formulation in "Earthwords."

56. Crimp, "On the Museum's Ruins," 47.

57. Eugenio Donato, "The Museum's Furnace: Notes toward a Contextual Reading of *Bouvard and Pécuchet*," *Textual Strategies: Perspectives in Post-Structuralist Criticism*, ed. J. V. Harari (London: Methuen, 1980), 213–238.

58. Ibid., 223.

59. Owens allies deconstruction with site specificity, comparing it with Rauschenberg's art, which must "provisionally accept the terms and conditions it sets out to expose" and "affirm what it set out to deny" (Owens, "Allegorical Impulse," vol. 2, 71). Similarly, he notes that Longo's work "participates of necessity in the activity it denounces" (vol. 2, 72) and that Sherman's inscribes its own "impossible complicity" with the mimicry denounced by her work (vol. 2, 79).

60. Ibid., vol. 2,71; vol. 2, 80.

61. Helga Geyer-Ryan, *Fables of Desire* (Cambridge: Polity Press, 1994), 199.

62. Buci-Glucksmann, *Baroque Reason: The Aesthetics of Modernity*, trans. Patrick Camiller (1984; repr. London: Sage, 1994), 49.

63. Ibid., 63.

64. Ibid., 71.

65. Geyer-Ryan, 198.

66. Buci-Glucksmann, *Baroque Reason*, 89.

67. De Man, "Rhetoric of Temporality," 195–196.

68. Cf. De Man, "Form and Intent," where de Man describes how the hermeneutic process is misrecognized. The critics (he here addresses formalist ones) misrecognize their own interpretation as belonging to their object of study.

69. The distinctions between metaphor, synecdoche, and metonymy are much debated. For a summary discussion of the "Metaphorical-Metonymical dyad" versus a "fourfold conception of the tropes," see Hayden White, *Metahistory: the Historical Imagination in Nineteenth-Century Europe* (Baltimore: The Johns Hopkins University Press, 1973), 31–33n13. Note the more "mechanical" or "quantitative" manner associated with the trope of metonymy, compared to the "organic" and "qualitative" connections of synecdoche.

70. Paul de Man, *The Resistance to Theory* (Manchester: Manchester University Press, 1986), 64.

71. For a range of assessments of de Man's wartime journalism, see Christopher Norris, *Paul de Man: Deconstruction and the Critique of Aesthetic Ideology* (London: Routledge, 1988); Werner Hamacher, Neil Hertz, and Thomas Keenan, eds., *Responses: On Paul de Man's Wartime Journalism* (Lincoln: University of Nebraska Press, 1989).

72. De Man, "Literary History and Literary Modernity," *Blindness and Insight*, 152.

73. De Man, "Form and Intent in the American New Criticism," *Blindness and Insight*, 20–35 (an earlier version of the essay was published in French in 1966). De Man cites Auerbach on 23.

74. De Man, "The Dead-End of Formalist Criticism," *Blindness and Insight*, 229–245.

75. De Man, "Rhetoric of Temporality," 220.

76. De Man, "Dead-End of Formalist Criticism," 241.

77. Benjamin, *Trauerspiel*, 163–166; de Man, "Rhetoric of Temporality," 222.

78. De Man, "Dead-End of Formalist Criticism," 242.

79. De Man, "Rhetoric of Temporality," 207.

80. Ibid., 207, 222.

81. Ibid., 208.

82. Melville, "Notes on the Re-emergence of Allegory," 56, 90.

83. Ibid., 82.

84. Ibid., 91.

85. Fred Orton, *Figuring Jasper Johns* (London: Reaktion, 1994), 12–13.

86. Ibid., 146.

87. Ibid., 131.

88. Quilligan, *Language of Allegory*, 53.

89. Ibid., 63.

90. Fletcher, *Allegory*, 220–221.

91. Honig, *Dark Conceit*, 125.

92. Quilligan, *Language of Allegory*, 235, 241; Fletcher, *Allegory*, 219.

93. C. S. Lewis cited Fletcher, *Allegory*, 36. Ferry cited Quilligan, *Language of Allegory*, 179.

94. Fletcher, *Allegory*, 42–43.

95. Quilligan, *Language of Allegory*, 27, 29.

96. Ibid., 33. (Cf. Buci-Glucksmann on the woven, labyrinthine nature of Benjamin's conception of allegory.)

97. Ibid., 235.

98. Ibid., 277, 252. Another way of putting this is that allegory internalizes allegoresis. Quilligan argues that allegories are critical reflections on their own "threshold texts" (in contrast to privileging the role of an external textual authority).

99. De Man, "Dead-End of Formalist Criticism," 242.

100. As part of a more recent departure from the anti-Hegelianisms associated with the canonical phase of postmodern thought, theorists as different as Slavoj Žižek and Jean-Luc Nancy again sought to recover the emancipatory potential of Hegel's "tarrying with the negative." Slavoj Žižek, *Tarrying with the Negative: Kant, Hegel, and the Critique of Ideology* (Durham: Duke University Press, 1993); Jean-Luc Nancy, *Hegel and the Restlessness of the Negative* (Minneapolis: University of Minnesota Press, 2002). On Left-Hegelians, see also Jürgen Habermas, "Three Perspectives: Left Hegelians, Right Hegelians, and Nietzsche," in *The Philosophical Discourse of Modernity: Twelve Lectures* (Cambridge: Polity Press, 1987); and *The Young Hegelians: An Anthology*, ed. Lawrence S. Stepelevich (Cambridge: Cambridge University Press, 1983).

101. De Man, "Rhetoric of Temporality," 207.

102. Benjamin, "Central Park," 191.

103. Ibid., 165, 172, 178. For further discussion, see Christine Buci-Glucksmann, *Baroque Reason*, 79–80, 106–109.

104. Walter Benjamin, "On the Concept of History" (written 1940 and published in *GS*), *Selected Writings, vol. 4*, trans. Howard Zohn, 389–400. Some recent revivals of the Benjaminian "leap," which draw on these theses, include Daniel Bensaïd, *Marx for Our Times: Adventures and Misadventures of a Critique*, trans. Gregory Elliott (1995; London: Verso, 2002) and Michael Löwy, *Fire Alarm: Reading Walter Benjamin's On the Concept of History*, trans. Chris Turner (2001; London: Verso, 2005).

105. Benjamin, *Trauerspiel*, 183.

106. Ibid., 181–183.

107. Ibid., 81.

108. Ibid., 208, 198.

109. Ibid., 193, 195, 191, 194.

110. Ibid., 194, 175, 191. Cf. Adorno's essay "The Natural History of the Theatre" (1931–1933), *Quasi una Fantasia: Essays on Modern Music*, trans. Rodney Livingstone (London: Verso, 1998), 65–78.

111. Ibid., 187, 195.

112. Benjamin, "Central Park," 169.

113. Benjamin, *Trauerspiel*, 160.

114. Ibid., 175.

115. Ibid., 195.

116. Ibid., 186–187.

117. Ludwig Klages cited in Julian Roberts, *Walter Benjamin* (London and Basingstoke: Macmillan, 1982), 107.

118. Benjamin, *Trauerspiel*, 176.

119. Ibid., 177.

120. Carl Horst cited in Benjamin, *Trauerspiel*, 177. Horst's description, which concerns the plastic arts, and which is concerned with art's hybridization with rhetoric, parallels the aesthetic preferences later advanced by Greenberg and Fried.

121. Gillian Rose, *Dialectic of Nihilism: Post-Structuralisn and Law* (Oxford: Basil Blackwell, 1984), 4.

122. Benjamin, *Trauerspiel*, 176–177.

123. Benjamin, "Paris, the Capital of the Nineteenth Century" (1935), trans. Howard Eiland, *Selected Writings, vol. 3, 1935–1938* (Cambridge, MA: Belknap/Harvard, 2002), 40. Although unpublished in Benjamin's lifetime, the essay formed the center of a debate with Adorno.

124. Benjamin, *Trauerspiel*, 53–54.

125. Ibid., 56.

126. Ibid., 182. Cf. Siegfried Kracauer's late work, a historiographical study in which Kracauer considers the relation between historical contingency and philosophy, drawing on the metaphor of photography to challenge philosophical complacency and the dismissal of "particularity." Siegfried Kracauer: *History: The Last Things Before the Last* (Princeton: Markus Wiener, 1995); first published posthumously, and completed by Paul Oskar Kristeller, in 1969. Kracauer's earlier essay on photography—"a secretion of the capitalist mode of production"—draws out similar themes on the philosophy of history, presenting photography as the last (latest) stage of pictorial representation and situating representation in terms of symbol and allegory. In one section, he attempts to describe the separation between consciousness and historical contingency. Siegfried Kracauer, "Photography," *The Mass Ornament: Weimar Essays* (Cambridge, MA: Harvard University Press, 1995), 47–63; the collection, published in 1963, drew together essays Kracauer had written for the *Frankfurter Zeitung* and was dedicated to Adorno.

127. Any proclaimed relation to Hegel—whether positive or negative—has to be situated historically, for it invariably plays out problems of its own moment rather than to Hegel's. The challenge to a progressivist philosophy of history in Benjamin's work, and particularly in his later writing, has a historical context of interwar political debates, which the writings on Benjamin then layer with their own investments, obliging us to take "Hegel" less as signifying a defined position than a field of debate. Today, it often turns on the understanding of the relation of Benjamin to Adorno, and the mediation of the former's work by the latter or by Rolf Tiedemann, and a further filtering through a hybrid of Althusserian Marxism or Deleuzean philosophy. The argument also touches on the fraught question of Adorno's relation to Benjamin. Bainard Cowan, for instance, claims that Adorno and, subsequently, Rolf Tiedemann were responsible for putting a Hegelian gloss on an essentially anti-Hegelian project, an issue addressed further by Susan Buck-Morss. See

Bainard Cowan, "Walter Benjamin's Theory of Allegory," *New German Critique* 22 (1981): 109–122; Susan Buck-Morss, *The Dialectics of Seeing: Walter Benjamin and the Arcades Project* (Cambridge, MA: MIT Press, 1989), especially the sixth chapter, "Historical Nature: Ruin," which deals extensively with the theme of allegory; Buck-Morss usefully discusses Adorno's 1932 essay "The Idea of Natural History" where he compares and contrasts the treatments of the idea of "second nature" in Lukács and Benjamin, 160. Of the accounts that distance Benjamin from Hegel, by far the most impressive is that provided by Howard Caygill in his book *Walter Benjamin: The Colour of Experience* (London: Routledge, 1998). Caygill provides an original reading of Benjamin's oeuvre, seeing his thought as a critical recasting of Kant's concept of experience, which draws centrally on the method of active nihilism outlined by Nietzsche. Although I pursue a reading more sympathetic to Hegelian figurations and language and while I think that Caygill's account bends the stick too far away from these, his emphasis on the place of transitivity of Benjamin's mode of thought and his careful articulation of Benjamin's relation to active nihilism has been very helpful.

128. Benjamin, *Trauerspiel*, 160, 166, 189, 174, 175.

129. Benjamin, "Central Park," 169.

130. Ibid., 160, 194.

131. Ibid., 189, 191.

132. Ibid., 192–193.

133. Ibid., 197.

134. Benjamin, "Paris," 40; "Central Park," 169, 167.

135. Benjamin, *Trauerspiel*, 183.

136. Ibid., 176.

137. Ibid., 183.

138. Ibid., 183.

139. Ibid., 79.

140. Ibid., 183.

141. Ibid., 201.

142. Ibid., 209. Cf. Adorno's description of high and low culture: the two exist as "torn halves of an integral freedom, to which however they do not add up." Theodor Adorno, Letter to Walter Benjamin, 18 March 1936, trans. Harry Zohn in *Aesthetics and Politics: Debates between Bloch, Lukacs, Brecht, Benjamin, Adorno* (1977; repr. London: Verso: 1980), 123.

143. Benjamin, *Trauerspiel*, 207, 208, 210.

144. Ibid., 194.

145. Ibid., 194, 175.

146. Ibid., 235.

147. Ibid., 201, 176.

148. Ibid., 161.

149. De Man, "Rhetoric of Temporality," 206.

150. Benjamin, *Trauerspiel*, 163.

151. Marcel Proust, *Remembrance of Things Past*, in 3 volumes, trans. C.K. Scott Moncrieff and Terrence Kilmartin (London: Penguin, 1983), 1: 88.

152. Paul de Man, "Reading (Proust)" (1972), in *Allegories of Reading: Figural Language in Rousseau, Nietzsche, Rilke, and Proust* (New Haven: Yale University Press, 1979), 57. Proust's discussion of Giotto is also addressed by J. Theodore Johnson, Jr., "Proust and Giotto: Foundations for an Allegorical Interpretation of *A la recherche du temps perdu*," in *Marcel Proust: A Critical Panorama*, ed. Larkin B. Price (Urbana: University of Illinois Press, 1973), 168–205. De Man notes that his own understanding of allegory and Johnson's "have little in common" (de Man, "Reading (Proust)," 75n20).

153. De Man, "Reading (Proust)," 67.

154. Jameson, *Postmodernism*, 227.

155. Gérard Genette, "Métonymie chez Proust," *Figures III* (Paris: Editions de Seuil, 1972) and translated as *Figures*, 3 vols., (New York: Columbia University Press, 1982); Gilles Deleuze, "Antilogos," in *Proust et les signes* (Paris: Presses Universitaires de France, 1964) and translated by Richard Howard, *Proust and Signs* (New York: George Brazillier, 1972). Also cf. de Man's argument to Melville's on the "anxiety for the recovery of a new beyond" (Melville, "Notes on the Re-emergence," 64).

156. De Man, "Reading (Proust)," 72.

157. Ibid.

158. Ibid., 77.

159. Norris, *Paul de Man*, 61.

160. De Man, "Reading (Proust)," 75. Proust was the French translator of Ruskin; Benjamin the German translator of Proust.

161. John Ruskin, *Fors Clavigera: Letters to the Workmen and Labourers of Great Britain* (Letter 7 July 1871) in volume 27 of *The Library Edition of the Works of John Ruskin*, in 39 volumes, eds. E. T. Cook and Alexander Wedderburn (London: George Allen, 1903–1912), 130. Ruskin's amendment is given in an author's note added to the first two volumes of *Fors Clavigera*.

162. For Fletcher, it is precisely the mechanical qualities of allegory that are its virtue.

163. Proust, *Remembrance*, 88.

164. Ibid., 89.

165. *Charity* is devoid of charity because, as Francoise, she is quite literally an uncharitable sort or, as the kitchen maid, she is someone in need of charity. As de Man remarks: "the literal sense of this allegory treats its proper sense in a most uncharitable manner"; but just as Marcel will never dispense with Françoise, so Proust will never cast away "the thematic powers of literal representation"; indeed, de Man insists, he "would not be able to do so if he tried."

166. De Man, "Reading (Proust)," 75.

167. Ibid.

168. Stephen Melville notes that the turn to allegory is "a belated rewriting of 'irony'" that allows for broader application (Melville, "Notes on the Reemergence," 60).

169. De Man, "Rhetoric of Temporality," 226.

170. Ibid., 226.

171. Ibid., 213–214.

172. Fletcher, *Allegory*, 157, 159, 341.

173. Honig, *Dark Conceit*, 68.

174. Clifford, *Transformations of Allegory*, 44.

175. Quilligan, *Language of Allegory*, 68.

176. Ibid., 200.

177. Ibid., 104.

178. In full ironic consciousness, as Schlegel put it, there is only "permanent parabasis." Schlegel is unusual among the writers of the Romantic period in seeing allegory as central to the poetic. (See Buci-Glucksmann, *Baroque Reason*, 69).

179. De Man, "Rhetoric of Temporality," 222.

180. Ibid.

181. Norris contrasts Marxism's understanding of the unhappy consciousness as a historical predicament with de Man's sense of it marking an ontological gulf "between phenomenal and semantic orders of sense" which cannot be overcome or transcended. Norris suggests that while the figure is used primarily in de Man's early, more existentially informed writing, it becomes generalized in the later work (Norris, *Paul de Man*, 3, xix).

182. De Man, "Form and Intent," 34.

183. De Man, "The Rhetoric of Temporality," 223.

184. Walter Benjamin, entry for 18 January 1926, *Moscow Diary*, ed. Garry Smith, trans. Richard Sieburth (Cambridge, MA: Harvard University Press, 1986), 95. For a discussion of the centrality of Proust's discussion for Benjamin's conception of allegory, and of Benjamin's study for Adorno, see Robert Hullot-Kentor, *Things Beyond Resemblance: Collected Essays on Theodor W. Adorno* (New York: Columbia University Press, 2006), 88–90.

185. Benjamin, "Central Park," 186.

186. Buci-Glucksmann, *Baroque Reason*, 76–77.

187. Benjamin, *Trauerspiel*, 198. (Cf. Ibid., 192.); Herbert Cysarz cited Benjamin, *Trauerspiel*, 199.

188. Benjamin, *Trauerspiel*, 200, 201.

189. Ibid., 173. Manfredo Tafuri, *Architecture and Utopia*, 14.

190. Benjamin, *Trauerspiel*, 199.

191. Ibid., 206. Cf. Benjamin's comments in "Epistemo-Critical Prologue" where he compares the Baroque with decadence—and its exaggerated

language—with Aloïs Riegl's analysis of late antiquity, and his emphasis on "artistic will" or *Kunstwollen* (ibid., 54–55). Cf. also Benjamin's approval of Baudelaire because "He managed to devalue certain poetic liberties of the Romantics through his classical deployment of the alexandrine, and to devalue classicist poetics through the characteristic ruptures and defects he introduced into classical verse" (Benjamin, "Central Park," 168).

192. Quilligan, *Language of Allegory*, 42.

193. Ibid. 219. Cf. her comments on Pynchon and Nabokov.

194. Ibid., 221.

195. Ibid., 278.

196. Ibid., 64.

197. Ibid., 80, 70, 79.

198. Ibid., 235.

199. Ibid., 277, 241, 265.

200. Ibid., 278.

201. Benjamin, *Trauerspiel*, 232.

202. G. W. F. Hegel, *Aesthetics: Lectures on Fine Arts*, vol. 1, trans. T. M. Knox (Oxford: Clarendon Press, 1975/1988), 64, 66.

203. Ibid., 66.

204. Ibid., 67.

205. Hegel, *Phenomenology*, 132.

206. Ibid., 476.

207. Ibid., 468.

208. Ibid.

209. Ibid., 472.

210. Georg Lukács citing Schlegel, in *Young Hegel: Studies in the Relations between Dialectics and Economics*, trans. Rodney Livingstone (1947–1948; repr. London: Merlin Press, 1975), 462.

211. Benjamin, *Trauerspiel*, 230.

212. Ibid., 232.

213. See, for example, Benjamin, "Central Park," 191; "Karl Kraus" (1931) and "The Destructive Character" (1931), both trans. Edmund Jephcott, *Selected Writings, vol. 2, 1927–1934* (Cambridge, MA: The Belknap Press of Harvard University Press, 1999). Buci-Glucksmann links Benjamin's "destructive principle" with the need to make leaps across the "unbridgeable gulf"—which, she believes, Benjamin saw as separating Jewish history and messianic freedom—and consonant with a politics of anarchism and nihilism (Buci-Glucksmann, *Baroque Reason*, 63–64).

214. Walter Benjamin, "Left-Wing Melancholy" (1931), trans. Ben Brewster, *Selected Writings, vol. 2, 1927–1934*, 425. This essay was translated for *Screen* in Summer 1974.

215. Honig, *Dark Conceit*, 68, 179.

216. De Man, "Rhetoric of Temporality," 226.

217. Norris, *Paul de Man*, 78, 94. Timothy Bahti advances a similar argument in "Lessons of Remembering and Forgetting," *Reading de Man Reading*, ed. Lindsay Waters and Wlad Godzich (Minneapolis: University of Minnesota Press, 1989), 257.

218. Andrzej Warminski, "Introduction: Allegories of Reference," *Aesthetic Ideology: Paul de Man*, ed. Andrzej Warminski (Minneapolis: University of Minnesota Press, 1996).

219. Jameson, *Postmodernism*, 225.

220. Ibid., 225, 227. Jameson believes that de Man's "positions and the arguments are 'postmodern' . . . even if the conclusions are not" (255).

221. Ibid., 218–220.

222. Ibid., 220. Jameson also sees the errors of aesthetic ideology as necessary: "metaphor is and is not an 'error': it generates illusions; yet insofar as it is inescapable and part of the very fabric of language itself, 'error' does not seem a particularly suitable word for it" (228).

223. Ibid., 229, 233, 241.

224. Ibid., 227.

225. Ibid., 238.

226. Honig, *Dark Conceit*, 30, 7, 25.

227. T. J. Clark, *Farewell to an Idea: Episodes from a History of Modernism* (New Haven: Yale University Press, 1999), 166; originally published as "Freud's Cézanne," *Representations* 52 (1995): 94–122, citations from 117 and 116.

228. Paul de Man, "Phenomenality and Materiality in Kant," in *Hermeneutics: Questions and Prospects*, ed. Gary Shapiro and Alan Sica (Amherst: University of Massachusetts Press, 1984), 121–144. Orton's engagement with this essay remains unpublished, but for related argument, see Fred Orton, "(Painting) Out of Time," *Parallax* 3 (1996): 99–112.

229. Clark, *Farewell*, 165.

230. De Man, "Heidegger's Exegesis of Hölderlin," in *Blindness and Insight*, 265.

231. De Man, "Sign and Symbol in Hegel's *Aesthetics*," *Critical Inquiry* 8 (1982): 761-775. See also "Hegel on the Sublime," *Displacement: Derrida and After*, ed. Mark Krupnick (Bloomington: Indiana University Press, 1983), 139–153.

232. De Man, "Sign and Symbol," 771.

233. De Man's reading has been challenged by Raymond Geuss. See Raymond Geuss, "Critical Response I: A Response to Paul de Man" and Paul de Man, "Critical Response II: Reply to Raymond Geuss," *Critical Inquiry* 10 (1983): 375–390.

234. De Man, "Sign and Symbol," 773.

235. Ibid., 773, 771.

236. Ibid., 770.

237. Ibid., 773.

238. Ibid., 768, 774.

239. The closing paragraphs of the *Phenomenology* are also concerned with the "night." Adorno takes this as an example of Hegel's "juggling the meaning of 'Erinnerung'" (as he does, more functionally, with *aufheben*), seeing this "equivocation" as a critical technique which employs language ironically and through which "Hegel uses language to convict language of the empty pretense of its self-satisfied meaning." (Hegel, *Phenomenology*, 492; Adorno, *Hegel*, 115–116.)

240. Benjamin, *Trauerspiel*, 177.

241. Ibid., 175

242. Ibid., 232.

243. Benjamin, *Trauerspiel*, 166.

244. De Man, "Sign and Symbol," 775.

245. Buchloh, "Allegorical Procedures," 52.

246. Ibid., 44.

247. Ibid., 52.

248. Ibid., 52, 53.

249. Walter Benjamin, "The Author as Producer" (1934), trans. Edmund Jephcott, *Selected Writings, vol. 2*, 768–782.

250. Martha Rosler, "Notes on Quotes" (1982), *Decoys and Disruptions: Selected Writings, 1975–2001* (Cambridge, MA: MIT, 2004), 140. A short section of her essay had appeared in the context of another in 1981.

251. The critiques of expression and originality in modernism, then prominent, were probably also important contexts for Buchloh's attitude toward appropriation. Hal Foster, "The Expressive Fallacy," *Art in America*, January 1983, republished in *Recodings*. See also Krauss, *Originality of the Avant-Garde.*

252. Rosler, "Notes on Quotes," 144.

253. Crimp's open letter and the editors' reply appear as "To Our Readers," *October* 53 (1990): 110–112.

254. Benjamin, "Central Park," 187.

255. Hal Foster, *Design and Crime (and Other Diatribes)* (London: Verso, 2002), 127.

256. It was a few years later when Thomas Y. Levin published his essay on Benjamin and Riegl and a translation of the one of Riegl's essays. Thomas Y. Levin, "Walter Benjamin and the Theory of Art History," *October* 47 (1988): 77–83; Aloïs Riegl, "Rigorous Study of Art," 84–90.

4. The Immobilizations of Social Abstraction

1. Robert Smithson in conversation with Bruce Kurtz on April 22, 1972, *The Writings of Robert Smithson*, ed. Nancy Holt (New York: New York University Press, 1979), 202.

2. Karl Marx, *Grundrisse, Foundations of the Critique of Political Economy (Rough Draft)*, trans. Martin Nicolaus (1857/8; repr. London: Penguin, 1973), 162.

3. Fredric Jameson, "Postmodernism, or, the Cultural Logic of Late Capitalism," *New Left Review* 146 (1984): 87.

4. Hal Foster, *Recodings: Art, Spectacle, Cultural Politics* (Seattle: Bay Press, 1985), 171, 167.

5. Ibid., 148, 147 (first published as "For a Concept of the Political in Contemporary Art," *Art in America*, April 1984). Foster qualifies his point; his account will be addressed in more detail later.

6. See, especially, Hans Magnus Enzensberger, "The Aporias of the Avant-garde," in *The Consciousness Industry: On Literature, Politics and the Media* (1962; repr. New York: The Seabury Press, 1974); Harold Rosenberg, *The Tradition of the New* (1959; repr. London: Paladin, 1970); "Collective, Ideological, Combative," *The Avant-Garde*, ed. Thomas B. Hess and John Ashbery (New York: The Macmillan Company, 1968), 74–78; "D.M.Z. Vanguardism," in *The De-Definition of Art* (Chicago: University of Chicago Press, 1972), 217–218.

7. Peter Bürger, *Theory of the Avant-Garde*, trans. Michael Shaw (1974; repr. Minneapolis: University of Minnesota Press, 1984); Andreas Huyssen, "The Search for Tradition: Avant-Garde and Postmodernism in the 1970s," *New German Critique* 22 (1982): 223–240, and "Mapping the Postmodern," *New German Critique* 33 (1984): 5–52, both republished in *After the Great Divide: Modernism, Mass Culture and Postmodernism* (London: Macmillan, 1986).

8. Lucy R. Lippard and John Chandler, "The Dematerialization of Art," *Art International* 12, no. 2 (1968): 31–36; Lucy Lippard, "Preface" and "Postface," *Six Years: The Dematerialization of the Art Object* (London: Studio Vista, 1973), 263–264. On the challenge to the idea of "dematerialization": Terry Atkinson's letter (written for the group around *Art-Language*) to Lucy Lippard of March 23, 1968, "Concerning the article 'the dematerialization of art,'" abbreviated in *Six Years*, 43–44, and published in full by Alexander Alberro and Blake Stimpson, eds., *Conceptual Art: A Critical Anthology* (Cambridge, MA: MIT, 1999), 52–58. For the criticisms of depoliticization: Michael Claura and Seth Siegelaub, "L'art conceptual" (1973), in Alberro and Stimpson, 286–290; *The Fox* (1975); Ian Burn, "The Art Market: Affluence and Degradation," *Artforum* 13, no. 8 (1975): 34–37, and "The 'Sixties: Crisis and Aftermath (or the Memoirs of an Ex-Conceptual Artist)," *Art & Text* 1, no. 1 (1981): 49–65 both republished in Alberro and Stimpson, 320–333 and 392–408.

9. Andre Gorz, *Farewell to the Working Class: An Essay on Post-Industrial Socialism* (1980; repr. London: Pluto Press, 1982); Nicos Poulantzas, *Political Power and Social Classes*, trans. Timothy O'Hagan (1968; repr. London: Verso, 1978) and *Classes in Contemporary Capitalism*, trans. David Fernbach (1974; repr. London: Verso, 1978); Ernesto Laclau and Chantal Mouffe, *Hegemony and Socialist Strategy: Towards a Radical Democratic Politics* (London: Verso, 1985). For a critical assessment of this phenomenon, see Ellen Meiksins Wood, *The Retreat From Class: A New "True" Socialism* (London: Verso, 1986).

10. W. F. Haug, *Critique of Commodity Aesthetics: Appearance, Sexuality and Advertising in Capitalist Society* (1971; repr. Cambridge: Polity Press, 1986). Haug coined the term in 1963.

11. Rosalind E. Krauss, *Picasso Papers* (London: Thames and Hudson, 1998); Rosalind Krauss, *A Voyage on the North Sea: Art in the Age of the Post-Medium Condition* (London: Thames and Hudson, 2001), 15.

12. Joseph Leo Koerner and Lisbet Koerner, "Value," *Critical Terms for Art History*, ed. Robert S. Nelson and Richard Shiff (Chicago: University of Chicago Press, 1996), 300–307.

13. Indicative of *October*'s established presence as an institution is the parody *November* √-1 (Winter 2006).

14. Buchloh, "Conceptual Art 1962–1969: From the Aesthetic of Administration to the Critique of Institutions," *October* 55 (1990): 105–143.

15. Buchloh, "Theorizing the Avant-Garde," *Art in America* 72 (1984): 19–21; "Introduction," in *Neo-Avantgarde and Culture Industry: Essays on European and American Art from 1955–1975* (Cambridge, MA: MIT, 2000), xxiv.

16. Buchloh, "Introduction," xxv.

17. Buchloh, "Roundtable: The Predicament of Contemporary Art," *Art Since 1900: Modernism, Antimodernism, Postmodernism*, Hal Foster et al. (London: Thames and Hudson, 2004), 672–673.

18. Buchloh, "Hantaï/Villeglé and the Dialectics of Painting's Dispersal" (1999), in *Neo-Avantgarde*, 254.

19. Buchloh appended an "Epilogue on the Idea of Use Value" to his essay "Moments of History in the Work of Dan Graham" (1978), in *Neo-Avantgarde*, 197–199.

20. Buchloh, "The Museum and the Monument: Daniel Buren's *Les couleurs/Les formes*" (1981), in *Neo-Avantgarde*, 126.

21. Buchloh, "Michael Asher and the Conclusion of Modernist Sculpture" (1980), in *Neo-Avantgarde*, 31.

22. Buchloh, "Museum and Monument," 133.

23. Ibid., 132.

24. Buchloh, "Moments of History," 188; and "Museum and the Monument," 132.

25. Buchloh, "Museum and Monument," 132.

26. Buchloh, "Parody and Appropriation in Francis Picabia, Pop, and Sigmar Polke" (1982), in *Neo-Avantgarde*, 348.

27. Ibid., 348.

28. Buchloh, "Moments of History," 188.

29. Buchloh, "Allegorical Procedures: Appropriation and Montage in Contemporary Art," *Artforum* 21, no. 1 (1982): 44.

30. Buchloh, "Moments of History," 198.

31. Ibid., 199, 191; Haacke cited by Buchloh, "Hans Haacke," 221, and ibid., 220. See also "Michael Asher"; "Museum and the Monument"; "Hans Haacke: Memory and Instrumental Reason" (1988), in *Neo-Avantgarde*, 203–241.

32. Buchloh, "Parody and Appropriation," 349.

33. Buchloh, "Moments of History," 188.

34. Buchloh, "Hans Haacke," 226.

35. Ibid., 222.

36. Buchloh, "Moments of History," 198.

37. Buchloh, "Michael Asher," 30.

38. Buchloh, "Moments of History," 198. See also Buchloh, "The Primary Colors for the Second Time: A Paradigm Repetition of the Neo-Avant-Garde," *October* 37 (1986): 40–52.

39. Buchloh, "Parody and Appropriation," 351.

40. Buchloh, "Introduction," xxiv. Buchloh has also defined the period of the neo-avant-garde slightly differently as 1958–1968 in "Plenty or Nothing: From Yve Klein's *Le vide* to Arman's *Le plein*" (1998), in *Neo-Avantgarde*, 262.

41. Buchloh, "Robert Watts: Animate Objects, Inanimate Subjects" (2000), in *Neo-Avantgarde*, 535.

42. Buchloh, "Robert Watts," 536, 535.

43. Buchloh, "Structure, Sign, and Reference in the Work of David Lamelas" (1997), in *Neo-Avantgarde*, 335. Note Buchloh's other essays on the mnemonic function: "Hans Haacke"; "Plenty or Nothing"; and "Memory Lessons and History Tableaux: James Coleman's Archaeology of Spectacle" (1995), *Neo-Avant-garde*, 141–178.

44. Buchloh, "Plenty or Nothing," 260.

45. Foster, *Art Since 1900*, 671–679.

46. Hal Foster, *The Anti-Aesthetic: Essays on Postmodern Culture* (Port Townsend, WA: Bay Press, 1983), published in the UK under the title *Postmodern Culture* (London and Sydney: Pluto Press, 1985); *Recodings: Art, Spectacle, Cultural Politics* (Seattle: Bay Press, 1985).

47. Hal Foster, "Postmodernism: A Preface," *Postmodern Culture*, ix–xvi; see especially xii; "(Post)Modern Polemics" (1984), *Recodings*, 121–136. Cf. Benjamin H. D. Buchloh, "Figures of Authority, Ciphers of Regression: Notes on the Return of Representation in European Painting," *Modernism and Modernity*,

ed. Benjamin H. D. Buchloh and Serge Guilbaut (Halifax, Nova Scotia: Nova Scotia College of Art and Design Press, 1983) and in *October* 16 (1981).

48. Foster, "Postmodernism: A Preface," xii.

49. Foster, *Recodings*, 139–155 and 157–179.

50. Ibid., 168.

51. Theodor Adorno and Max Horkheimer, *Dialectic of Enlightenment*, trans. John Cumming (1944; repr. London: Verso, 1986), 7. See also Max Horkheimer, *Eclipse of Reason* (1947; repr. London: Continuum, 2004).

52. It would be instructive to compare and contrast Foster's writings with the work of British art theorist John Roberts at this time. See John Roberts, *Postmodernism, Politics and Art* (Manchester: Manchester University Press, 1990); *Selected Errors: Writings on Art and Politics 1981–90* (London and Boulder, Colorado: Pluto, 1992).

53. Foster, *Recodings*, 132.

54. Ibid., 130.

55. Foster, "Whatever Happened to Postmodernism?" in *Return of the Real: The Avant-Garde at the End of the Twentieth Century* (Cambridge, MA: MIT, 1996), 217. An earlier version of this chapter was "Postmodernism in Parallax," *October* 63 (1993): 3–20 (Winter 1993).

56. Hal Foster, *Design and Crime (and Other Diatribes)* (London: Verso, 2002), 130.

57. Ibid., 52. This chapter, "Architecture and Empire," first appeared in the *London Review of Books*. (For Foster, this type of "reprogramming" should be distinguished from the "retooled and recycled" approach to style that he identified with neoconservative postmodernism. See, *Recodings*, 121.)

58. Foster, *Design and Crime*, 60.

59. Ibid., 61.

60. Ibid., 59n19; 61.

61. Ibid., 129.

62. Rosalind E. Krauss, "Sculpture in the Expanded Field," *October* 8 (1979).

63. Foster, *Design and Crime*, 127.

64. Rem Koolhaas cited in *Design and Crime*, 62.

65. Foster, *Return of the Real*, 205–207.

66. Jameson, "Postmodernism," 78.

67. Foster, *Recodings*, 144–145.

68. Foster, *Return of the Real*, 207.

69. Note Foster's earlier comments in "Postmodernism: A Preface": there is not a clean break between modernism and postmodernism, he argues, but rather uneven developments. In his introduction to *Recodings*, he counters the suggestion that postmodernism represents a historical rupture with the modern.

70. Jameson, *Postmodernism*, xxii.

71. Jameson, *A Singular Modernity: Essay on the Ontology of the Present* (London: Verso, 2002), 13. Terms such as "modern"—and, by extension, "postmodern"—Jameson also argues, are problem *words*, but do not acquire the status of concepts.

72. Jameson, "Postmodernism." See also Jameson's comments on the demise of parody and loss of coordinates, 64–65.

73. Jameson, *Postmodernism*, xix.

74. Foster's discussion of the mnemonic is in *Design and Crime*, 130.

75. Jameson, "Postmodernism," 66.

76. Foster, *Art Since 1900*, 587.

77. The French translation of *Das Kapital* makes the identification with Saussure's theory a relatively smooth one. (Thanks to Tom Gretton for drawing my attention to this.)

78. Jean Baudrillard, *The Mirror of Production*, trans. Mark Poster (1973; repr. St. Louis: Telos Press, 1975), 23, 24.

79. Roman Rosdolsky, *The Making of Marx's Capital*, trans. Pete Burgess (London: Pluto Press: 1977), 73. Paul M. Sweezy, *The Theory of Capitalist Development* (1942; repr. New York: Modern Reader Paperbacks, 1970).

80. Marx, *Grundrisse*, 646–647.

81. Jean-Joseph Goux, "Marx et l'inscription du travail," *Tel Quel* 33 (1968), translated in *The Tel Quel Reader*, ed. Patrick ffrench and Roland-François Lack (London: Routledge, 1998), 50–67.

82. Jameson, *Brecht and Method* (London: Verso, 1998), 50; cf. Jameson, *Postmodernism*, 187.

83. Jameson, *Postmodernism*, 181ff. See, for example, the work assembled in Martha Woodmansee and Mark Osteen, eds., *The New Economic Criticism: Studies at the Intersection of Literature and Economics* (London: Routledge, 1999).

84. Theodor Adorno, Letter to Walter Benjamin, 2 August 1935, Ernst Bloch et al., *Aesthetics and Politics* (London: New Left Books, 1977), 113.

85. Theodor W. Adorno, "On the Fetish-Character in Music and the Regression of Listening" (1938), *The Essential Frankfurt School Reader* (Oxford: Basil Blackwell, 1978), 279.

86. Theodor Adorno, *Aesthetic Theory*, trans. Robert Hullot-Kentor (1970; repr. London: Athlone Press, 1997), 17, 21.

87. Adorno, *Aesthetic Theory*, 21–22. Cf. Herbert Marcuse, *One-Dimensional Man: The Ideology of Industrial Society* (London: Sphere Books, 1968); Enzensberger's account of the consciousness industry.

88. Peter Dews, *Logics of Disintegration: Post-structuralist Thought and the Claims of Critical Theory* (London: Verso, 1987), 150.

89. Adorno, *Aesthetic Theory*, 227, 17.

90. Ibid., 227.

91. Guy Debord, *The Society of the Spectacle*, trans. Donald Nicholson-Smith (1967; repr. New York: Zone Books, 1994), 44, 45, 32, 33.

92. Ibid., 32.

93. Ibid., 14

94. Ibid., 32, 14.

95. Ibid., 22, 23.

96. Ibid., 143–144.

97. Ibid., 24.

98. Ibid., 12, my emphasis.

99. Ibid., 19.

100. Ibid., 146–147.

101. Jameson, *Postmodernism*, 18.

102. Marx, *Capital*, vol. 1, 78.

103. See, for example, Agnes Heller, *The Theory of Need in Marx* (London: Allison and Busby, 1976); Ian Fraser, *Hegel and Marx: The Concept of Need* (Edinburgh: Edinburgh University Press, 1998).

104. Adam Smith, *The Theory of Moral Sentiments*, Part 4 (London: A. Miller, 1759 and Edinburgh: A. Kincaid & J. Bell, 1759).

105. Nicholas Barbon, *A Discourse Concerning Coining the New Money Lighter: In answer to Mr Lock's Considerations about Raising the Value of Money* (London: Richard Chiswell, 1696), 2, 3; Nicholas Barbon, *A Discourse of Trade* (London: Tho. Milbourn, 1690), 13–15.

106. We should also note the extent to which aesthetic issues are located in these same discussions.

107. István Mészáros, *Beyond Capital: Towards a Theory of Transition* (London: Merlin, 1995).

108. Marx, "Preface to First German Edition" (1867), *Capital*, vol. 1, 18–19.

109. Revisions and extensions were also made to the opening section and to the section devoted to the fetishism of commodities in Chapter 1, to the first section of Chapter 3, and to the second section of the current Chapter 9. The Third and Fourth German editions were prepared by Engels, with the latter the basis for the English translation.

110. Marx to Engels (22 June 1867), Marx and Engels, *Selected Correspondence* (Moscow: Progress Publishers, 1955), 177; Marx, "Afterword to the Second German Edition" (1873), *Capital* vol. 1, 22. See also Marx to Engels (24 August 1867), *Selected Correspondence*, 180.

111. Marx, *Capital*, vol. 1, 28, "Afterword to the Second German Edition." (1873).

112. Ibid., vol. 1, 54.

113. Ibid.

114. Rubin usefully distinguishes the various "forms of value" (simple, expanded, general, and money forms) from the discussion of "value as form," the "value form" or the "'form of value' as such." Rubin, 112n1, 114.

115. Marx, *Capital,* vol. 1, 76–77; vol. 1, 54; vol. 1, 55.

116. Cf. Ibid., vol. 1, 56, vol. 1, 55.

117. Ibid., vol. 1, 67.

118. Ibid., vol. 1, 67.

119. Marx, *Grundrisse,* 148.

120. Ibid., 147.

121. Karl Marx, "Notes on Adolph Wagner's *Lehrbuch der politischen Ökonomie (Zweite Auflage), Band 1, 1879* (1881).

122. Cf. Rubin, 41–42. What and how presuppositions are made is a central feature of Marx's critique of the methods of classical political economy.

123. The "merely formal manner of presentation" appropriate to the first volume, Marx argued in Volume 2, was "no longer adequate" for the second (Marx, *Capital,* vol. 2, 356–357). Cf. Rosdolsky, 83–88.

124. Fredric Jameson, "Cognitive Mapping," *Marxism and the Interpretation of Culture,* ed. Cary Nelson and Lawrence Grossberg (Urbana and Chicago: University of Illinois Press, 1980), 347–360; *The Seeds of Time* (New York: Columbia University Press, 1994); *Archaeologies of the Future: The Desire Called Utopia and Other Science Fictions* (London: Verso, 2006).

125. Fredric Jameson, *Marxism and Form: Twentieth-Century Dialectical Theories of Literature* (Princeton: Princeton University Press, 1971), 306–416.

126. Jameson, "Postmodernism," 75.

127. Jameson, "Culture and Finance Capital," *The Cultural Turn,* 146.

128. Fredric Jameson, *The Prison-House of Language: A Critical Account of Structuralism and Russian Formalism* (Princeton: Princeton University Press, 1972); Jameson, *Postmodernism,* 181ff.

129. Jameson, "Culture and Finance Capital" (1998) and "The Brick and the Balloon: Architecture, Idealism and Land Speculation"(1997/8), *The Cultural Turn: Selected Writings on the Postmodern, 1983–1998* (London: Verso, 1998), 136–161 and 162–189.

130. Jameson, *Postmodernism,* 187.

131. Ibid., 187.

132. Jameson, "Brick and Balloon," 162–163.

133. Harvey argues that the strategies adopted by modern financial capital were already well perfected by landed property prior to the emergence of industrial capital; David Harvey, *The Limits to Capital* (Oxford: Basil Blackwell, 1982), 370. Rosdolsky, however, thinks it important to distinguish modern interest from ancient forms of usury, and to show the historically specific role of interest for capital. (Rosdolsky, *Making of Marx's Capital,* 386–389.)

134. Jameson, "Brick and the Balloon," 187. (Cf. *Postmodernism,* 185.)

135. Ibid., 185.

136. Ibid., 187. (Cf. *Postmodernism,* 399.)

137. For Harvey's discussion of "historical-geographical materialism," see *Justice, Nature and the Geography of Difference* (Oxford and Malden, MA: Basil

Blackwell, 1996); "Space as a Key Word," *Spaces of Global Capitalism* (London: Verso, 2006).

138. Jameson, "Brick and Balloon," 186.

139. Ibid.

140. Ibid.

141. Ibid., 186–187.

142. Ibid., 187.

143. Jameson, "Culture and Finance Capital," 160.

144. Ibid., 160–161.

145. Marx, *Capital*, vol. 3, 464–72.

146. Harvey, *Limits to Capital*, 269.

147. Marx, *Grundrisse*, 623.

148. Harvey, *Limits to Capital*, 288.

149. Ibid., 327.

150. Ibid., 269.

151. On the history of money and credit, see, for example, Ernest Mandel, *Marxist Economic Theory*, trans. Brian Pearce (1962; repr. London: Merlin Press, 1977); his key points are extended in *Late Capitalism*, trans. Joris De Bres (1972; repr. London: Verso, 1978).

152. Marx, *Grundrisse*, 115–139.

153. Ibid., 239–330.

Afterword: Abstract and Transitive Possibilities

1. Radek Community, *Manifestation* (video, 2001). Named after the left Bolshevik Karl Radek, the group was founded in 1997 and disbanded at the beginning of 2008.

2. *Le Souvenir d'un avenir* [*Remembrance of Things to Come*], video, directed by Chris Marker and Yannick Bellon (Brooklyn, NY: First Run/Icarus Films, 2001).

3. Theodor Adorno, "Extorted Reconciliation: On Georg Lukács's *Realism in Our Time*" (1958), *Notes to Literature*, in 2 volumes, ed. Rolf Tiedemann, trans. Shierry Weber Nicholsen (New York: Columbia University Press, 1991), vol. 1, 221. Focused through this extended review, Adorno's response otherwise appears as scattered passages in essays and papers from the period. The designations also refer to essays and published lectures by Lukács, such as "Healthy or Sick Art?" in *Writer and Critic and other essays*, trans. Arthur Kahn (London: Merlin, 1978).

4. Adorno, "Extorted Reconciliation," 234.

5. *Erpresste Versöhnung* is translated as "Reconciliation under Duress" by Rodney Livingstone for *Aesthetics and Politics*, Ernst Bloch et al. (London: New Left Books, 1977).

6. Georg Lukács, *The Young Hegel: Studies in the Relations between Dialectics and Economics*, trans. Rodney Livingstone (1954; repr. London: Merlin Press, 1975), xi. Completed in 1938, the book was first published in revised form in 1947/8, with the "new edition" appearing in 1954.

7. Ibid., 419.

8. Ibid.

9. Ibid., 419, 418.

10. Ibid., 404.

11. Theodor Adorno, "Skoteinos, or How to Read Hegel" (1962–1963), in *Hegel: Three Studies*, trans. Shierry Weber Nicholson (1963; repr. Cambridge, MA: MIT Press, 1993), 107–9.

12. Ibid., 147, 141.

13. Adorno, "Aspects of Hegel's Philosophy" (1956) and "The Experiential Content of Hegel's Philosophy" (1959), in *Hegel: Three Studies*, 51 and 83.

14. Adorno, "Skoteinos," 134; "Aspects," 10; "Skoteinos," 138.

15. Adorno, "Experiential Content," 62; "Aspects," 13.

16. Adorno, "Skoteinos," 142, my emphasis.

17. Ibid., 142, 143.

18. Lukacs, *Young Hegel*, xviii.

19. Ibid., xvi, xxi.

20. Ibid., xviii.

21. Adorno, "Aspects," 1.

22. See, for example, the journal *Actuel Marx*, or the coments made by Daniel Bensaïd in "Preface to the English Edition: The Archipelago of a Thousand Marxisms" (2002), in *Marx for Our Times: Adventures and Misadventures of a Critique* (1995: London: Verso, 2002), ix.

23. Adorno, "Aspects," 1.

24. Ibid.

25. Adorno, "Experiential Content," 80 (cf. "Skoteinos," 124), 70.

26. Adorno, "Skoteinos," 143.

27. Adorno, "Experiential Content," 82, 78.

28. The issue here also touches on one encountered earlier: Marx's comment in the first of his "Theses on Feuerbach" that it was idealism, rather than materialism which contributed the "active side." Adorno develops this theme with his emphasis on the "experiential substance" in Hegelian speculation (and in Fichte and Schelling). (60)

29. Lukács's *Realism in Our Time* appears in English as *The Meaning of Contemporary Realism*, trans. John and Necke Mander (1957; repr. London: Merlin, 1963). The discussion of potentiality is on 21ff.

30. Adorno, "Experiential Content," 84.

31. Adorno, "Skoteinos," 132.

32. Adorno's letter to Mann cited in "Experiential Content," 83–84.

33. Adorno, "Extorted Reconciliation," 222.

34. Ibid., 222–223. As an example of the opposite problem to the one he identifies in Lukács's approach, Adorno cites Brecht's characterization of Arturo Ui, where the caricature was so oversimplified that it became an "unrealistic device," and which, having failed to "characterize its object adequately" diminished the effectiveness of Brecht's satire (222).

35. Adorno, "Skoteinos," 136.

36. Ibid., 123.

37. Ibid., 122; "Experiential Content," 91; "Skoteinos," 121–3. Schelling may have set out to make "experience" the subject of his philosophy, Adorno argues, but it was Hegel who achieved it in practice by having us experience the concept in process so that "understanding has to find a foothold in the gap between experience and concept" ("Skoteinos," 138).

38. Adorno, "Experiential Content," 84.

39. Adorno, *Aesthetic Theory*, trans. Robert Hullot-Kentor (London: Athlone, 1999), 196.

40. Ibid., 192.

41. Georg Lukács, *The Theory of the Novel: A Historico-Philosophical Essay on the Forms of Great Epic Literature*, trans. Anna Bostock (1916; repr. Cambridge, MA: MIT, 1971), 78.

42. Karl Radek was a Left communist who at this point was secretary for the executive of the Communist International. He went on to support the Left Opposition from 1924–1929 and died in a Russian camp in 1939. The Radek Community appealed particularly to his activistic associations. The March Action focused a crisis over the balance between spontaneous resistance and the centralized control of the KPD. *Sollen* conveys not only "ought" but also ideas of hope and desire. For and extended discussion of the tragic view in Lukács's anticapitalism, see Michael Löwy, *Georg Lukács—From Romanticism to Bolshevism*, trans. Patrick Camiller (1976; repr. London: New Left Books, 1979).

43. István Mészáros, *Lukács' Concept of Dialectic* (London: Merlin Press, 1972), 93.

44. Mészáros., 43; Lukács cited in ibid., 44.

45. Ibid., 43.

46. T. J. Clark, "Jackson Pollock's Abstraction," *Reconstructing Modernism: Art in New York, Paris, and Montreal, 1945–1964*, ed. Serge Guilbaut (Cambridge, MA: MIT, 1990), 222.

47. Mario Tronti has reflected that the attempt by intellectuals associated with the *operaisti* to wrest negative thought from conservative philosophy was conceived as a Marxian gesture that intended to remake the Left's intellectual resources. The ambition was to develop a form of a modern realism that would address social totality not from the point of view of the whole (as did conservatives) but from the perspective of the part(isan).

48. Chto delat (What is to be Done) is a collective of artists, critics, philosophers, and writers founded in Petersburg in 2003. They describe their project as a "self-organizing platform for cultural workers intent on politicizing their 'knowledge production' through reflections and redefinitions of an engaged autonomy for cultural practice today." They publish a journal of the same name (available online).

Index

GPSR Authorized Representative: Easy Access System Europe, Mustamäe tee 50, 10621 Tallinn, Estonia, gpsr.requests@easproject.com

www.ingramcontent.com/pod-product-compliance
Ingram Content Group UK Ltd.
Pitfield, Milton Keynes, MK11 3LW, UK
UKHW041624260326
469231UK00007B/55/J
9780231149389